SOTHIS PRESS

Sothis Press aims to completely reimagine the public domain book with an emphasis on universal access to captivatingly designed print editions. We believe that technology will enable every single book ever published to be perpetually accessible in beautiful print form. Sothis is beauty in ubiquity.

www.sothis.press
1 Waterhouse Square, London, ECIN 2ST

The Agaves of Baja California

Howard Scott Gentry

SOTHIS PRESS

ISBN 978-1-0365-1470-9

This edition is published by
Sothis Press © 2023. The company
is registered in England and Wales.
The Company Number is 14900389.

OCCASIONAL PAPERS

OF THE

CALIFORNIA ACADEMY OF SCIENCES

No. 130, 119 pages, frontis. +6 pls., 62 figures, 11 tables — December 28, 1978

The Agaves of Baja California

By

Howard Scott Gentry

Desert Botanical Garden,
Papago Park, Phoenix, Arizona 85010

SAN FRANCISCO

PUBLISHED BY THE ACADEMY

The Agaves of Baja California

Frontispiece. The wonderful vegetation north of Punta Prieta with Sierra San Luis in the background. *Agave shawii goldmaniana*, *Yucca valida*, *Fouquieria splendens*, and *Idria columnaris* are all prominent here. Photographed, 21 February 1968.

OCCASIONAL PAPERS

OF THE

CALIFORNIA ACADEMY OF SCIENCES

No. 130, 119 pages, frontis. +6 pls., 62 figures, 11 tables December 28, 1978

The Agaves of Baja California

By

Howard Scott Gentry

Desert Botanical Garden,
Papago Park, Phoenix, Arizona 8501

SAN FRANCISCO

PUBLISHED BY THE ACADEMY

US ISSN 0068-5461

The California Academy of Sciences
Golden Gate Park
San Francisco, California 94118

PRINTED IN THE UNITED STATES OF AMERICA
BY ALLEN PRESS, INC., LAWRENCE, KANSAS

TABLE OF CONTENTS

LIST OF COLOR PLATES

ACKNOWLEDGMENTS

Of first mention is Carl O. Erlanson, former Chief of Plant Exploration and Introduction, Agricultural Research Service, who first assigned me to formulate a taxonomic and economic monograph of the genus *Agave*. I am grateful to him for such a challenging assignment. Others in that unique institution who assisted in my early efforts were Arthur S. Barclay, Frederick J. Hermann, Howard L. Hyland, Ernest Imle, Quentin Jones, Frederick G. Meyer, Bernice Schubert, and C. Earle Smith, Jr. Field assistants who were especially helpful in Baja California included Juan Arguelles of San Bernardo, Sonora, and Frank L. Cech of San Diego, California.

I am indebted to the fine artists Regina Hughes, Lucretia Hamilton, and Wendy Hodgson, whose drawings enhance this book. I wish also to thank Bruce Gentry, John McClure, and Reid Moran for the photographs they provided for illustrations and for study. Donald Pinkava and Reid Moran reviewed the script and suggested many changes, many of which I gratefully accepted. Thomas Howell has competently edited my Latin diagnoses.

The color illustrations in this book are due to the generous support of E. R. Le Roy of San Francisco, California. Furthermore, through correspondence, he has given the author interest and encouragement during the past decade.

Final efforts in field and laboratory were provided for by the National Science Foundation with their fund Grant No. BMS74-24553. With this and the facilities so magnanimously provided by the Desert Botanical Garden and Director W. Hubert Earle, I could do no less than my best to make this a useful and enduring work. Thank you all, including the many others not called by name!

HOWARD SCOTT GENTRY
June 5, 1975

ABSTRACT

Gentry, Howard S. The agaves of Baja California. *Occasional Papers of the California Academy of Sciences*, no. 130, 119 pages, frontis. + 62 figures, 11 tables, 1978.—This is a systematic and economic account of plants of the genus *Agave*, family Agavaceae, that are indigenous to Baja California and the adjacent areas of the Gulf of California. Taxonomically, it is a revision of the three generic groups established by William Trelease in 1912: Deserticolae, Campaniflorae, and Umbelliflorae. Of the 23 taxa described, 4 are proposed as new species: *Agave capensis*, *A. gigantensis*, *A. moranii*, and *A. vizcainoensis*. Eight other new taxa proposed are: *A. deserti* ssp. *pringlei*; *A. deserti* ssp. *simplex*; *A. cerulata* ssp. *dentiens*; *A. cerulata* ssp. *nelsonii*; *A. cerulata* ssp. *subcerulata*; *A. sobria* ssp. *frailensis*; *A. sobria* ssp. *roseana*; *A. shawii* ssp. *goldmaniana*. Keys to assist in identification are provided for groups, species, and subspecies. Maps show the distributions of all taxa and nearly all are copiously illustrated with line drawings and photographs.

The economic aspects of the plants are represented by historical notes on Indian uses. Under various species there are reports of modern attempts to exploit agaves for fiber and chemical constituents. Tables on food consumption, chemical analyses, and fiber tests summarize the information. Sustaining documentation is provided in a bibliography, a glossary of special terms, and extensive citations of the specimens reviewed in many U.S. herbaria. The latter have been organized in an appendix of exsiccatae listed alphabetically by species, states, and collectors with respective herbaria. This report is the culmination of a field study of the wild populations of these plants made intermittently by the author over a 30-year period.

INTRODUCTION

This work is primarily a taxonomic revision of three groups of the large genus *Agave* in the family Agavaceae: Deserticolae, Campaniflorae, and Umbelliflorae. Its scope has been enlarged beyond the taxonomic skeleton in order to include the historic uses of the plants and to report investigations of their chemistry and fiber. These investigations were made by the Agricultural Research Service of the U.S. Department of Agriculture, the procurement of the samples being done by the writer. The results are condensed in tabular form with brief textual comments. For the purpose of brevity, many special articles in the press and journals have not been repeated or reviewed here. A great part of this study is based on extensive observations of the wild agave populations. These observations form a basic part of the writer's taxonomic concepts, and for this reason, many field notes are given space in the accounts of species.

Trelease's (1912) account of the agaves of Lower California included descriptions of 23 species. It was based on a relatively small scattering of specimens from field collections made principally by J. N. Rose, T. S. Brandegee, E. W. Nelson, and E. A. Goldman. The present account recognizes 16 species, 3 of which are proposed as new, 8 subspecies, and 1 variety. Several taxa are outside the confines of Baja California, but they complete this revision of the group Deserticolae. Trelease arranged this species in groups. I regard these groups as equivalent to sections, but am not taxonomically formalizing them as such in this revision.

In addition to the three groups listed above, there remain two Californian taxa that Trelease put into his small group Datyliones. I am including them in this work also so that all known wild agaves of Baja California are accounted for. As further preamble, the environmental background is discussed below. The harmony that naturally exists between agave and other life forms of the peninsula is clearly apparent in the frontispiece.

ENVIRONMENT AND EVOLUTION

The California Gulf region is physiographically unique with its long slender peninsula separated from the Sierra Madrean mainland by a rather deep seaway. The geologic structures, especially the widespread volcanic formations, all indicate a tectonically active history during much of the Tertiary Period. The dynamic San Leandro fault system of upper California does not stop at the U.S.–Mexico border, but continues in the Gulf of California along the length of the peninsula. The escarpments on the eastern side of the peninsular mountain axis are the result of upthrusts along this fault zone, at least relative to the "graben" or trough where the sea lies.

Until recently the reading of the older geologic reports (as those of Darton 1921, and Schuchert 1935), left me with the physiographical impression that the present peninsula is actually a chain of land segments gradually united by volcanic magmas and land emergence. While this still appears true in most instances, modern students of marine geology, equipped with cunningly devised electronic equipment, are developing highly illuminative information about earth crustal movements. The proponents of tectonic plate movements (growing out of the continental drift hypothesis) now regard the peninsula as part of a land raft slipping northward along the western edge of the North American continent. *The Scientific American* with its "Continents Adrift" (Wilson et al. 1972) has put together a fine series of articles on this large subject. Other articles have been carried in *Science* during the 1960's. One recent study by Larson, Menard, and Smith (1968) concludes that southern Baja California was torn from the Mexican mainland about 4 million years ago and since has been rafted 260 km to the northwest at a rate of 3 cm per year. This is in conflict with earlier geologic studies that generally place the opening of the Gulf in the Miocene some ten to fifteen or more million years ago. While such discrepancies are being adjusted by further work, we can consider the possible effect of land movements on our plant subjects.

It would be presumptuous to try to correlate *Agave* evolution with land changes in the California Gulf region, but nevertheless, I will essay it in a suggestive and tentative way. Several of the *Agave* complexes are limited to definite land areas, and, in some cases, these land areas appear to have been separated by seaways during much of the upper Tertiary and Pleistocene periods. In Figure 1, I have drawn a map of peninsular land areas that were apparently separated from one another during much of these geologic periods. On these separated land areas

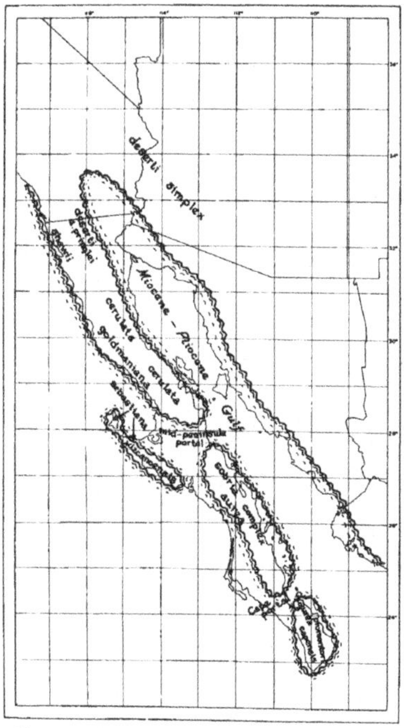

FIGURE 1. Physiographic staging and some agave distributions in the California Gulf region.

are written the respective agaves growing there today.

It all suggests that many of the distinctions among the species appear to have been fostered by the seaway separations. Starting in the north, various situations can be outlined as follows:

1. The *Agave deserti* complex was split into two groups by the Gulf: *A. deserti simplex* on the mainland northeast; *A deserti deserti* and *A. deserti pringlei* on the peninsular side.
2. *Agave sobria* is set off rather sharply morphologically and anatomically from *A. cerulata*, the two complexes also being separated by a postulated mid-peninsular sea portal.
3. *Agave vizcainoensis*, resident upon what was obviously a western island aligned with Cedros Island, with its deep floral tube, is quite distinct from any of its relatives in the Deserticolae.
4. *Agave margaritae, A. sobria roseana, A. cerulata dentiens* are all island endemics still limited, or nearly so, to their respective islands.
5. The Campaniflorae are limited to the southern part of the peninsula (south of the midpeninsular sea portal); *A. aurea* is widespread along the Sierra de la Giganta volcanics and is also scattered over the Cape District granitics. *Agave promontorii* and *A. capensis* are endemic to the Cape mountains. If their progenitors were originally based on the Mexican mainland 4 million years ago, one would expect to find near relatives along the Nayarit-Jaliscan coast, but no such relatives have been found there. The Campaniflorae are quite distinct from both mainland and peninsular agaves. This suggests long isolation, longer than the 4 million years allowed by Larson et al. (1968).

The overall picture suggests some coincidence between *Agave* evolution and land evolution and that many *Agave* species do not migrate readily. On the basis of fossil records for many flowering plant families, we can assume that agaves and their ancestors have been on the North American continent throughout the Tertiary Period. But without agavoid fossils this is uncertain. Nor do we really know how long agaves have inhabited the peninsula. Just a few fossil impressions of spines, seeds, or capsule sections would be of great historical value, and I hope the paleontologists will be looking for them.

Most of the California Gulf region has been well characterized as the Sonoran Desert (Shreve 1951), and the agaves living there are successful succulent xerophytes. Most of the desert species make do with only 90–250 mm (4–10 inches) of rain annually, and some of them, such as *A. deserti* and *A. cerulata*, can survive several years with little or no rainfall. *A. promontorii*, a large fleshy mesophytic species living on the Cape mountains, receives an annual average of ca. 750 mm (25 inches). *A. shawii*, alone among the agaves, occupies a unique type of climate in North America, the Mediterranean, with rainless summers and winter-spring rainfall. Its two close relatives southward along the outer coast are in drier latitudes, but frequent fogs ameliorate the drier condition of lower rainfall. Graphic details of rainfall patterns are provided in Figures 2, 44, 51, along with indications of the flowering seasons for each taxa.

Agaves are generally abundant in Baja California and in certain areas form a dominant element in the vegetation. *Agave shawii goldmaniana* is a green succulent conspicuous over many kilometers on the sandy uplands of the central peninsula (Fig. 3). *Agave cerulata* is also there by the thousands, but its small gray size and yellowish color keep it less conspicuous. These are easily seen along the main road, but many other species can be found only by side trips to the mountain slopes by car and afoot. Among these are *A. moranii* on the eastern slopes of the Sierra San Pedro Mártir, *A. gigantensis*, a relatively rare species of the Sierra de la Giganta, and still others can be reached only by boat to the peninsular islands.

Except during years of protracted drought, the tall masts of flowering agaves adorn the roadways from late winter to early summer. Yellow is the common color, but orange hues are frequent, while some species show bracts and buds of bright red or purple. The masts persist with dry and weathering capsules, sometimes for years, forming vibrant silhouettes on skylines. The varying forms of rosettes, leaves, and inflorescences are countless when observed closely and are confusing when one attempts to identify them to species. Even trained botanists have turned away from them in a kind of taxonomic despair. A main purpose of this work is to bring this complex pageant of plant life into an understandable and simplified perspective.

The taxonomic characters of *Agave* have been discussed elsewhere (Gentry 1972) and will not be recounted here. However, there is a growth habit which is quite striking in the *A. shawii* group. These plants are perennials, since they flower repeatedly, and are truly arborescent, developing elongate trunks. They bud and branch from the leaf axils and reach several meters in length. The trunk cannot support the heavy succulent weight and lies down upon the ground. New roots form with soil contact as the older leaves die off and rot away. The older rosettes flower and dry, and as old sections of stem deteriorate, the younger rosettes continue to grow, to branch, and to root until a fragmented reclining tree or clone is formed (Fig. 3). Large clones may cover many square yards with the appearance of many individual plants. Such clones appear to be several hundred years old.

TABLE 1. SUMMARY OF FOODS USED BY INDIANS IN TH CENTRAL DESERT REGION OF BAJA CALIFORNIA. (Fror Aschmann 1959:103–104.)

Typical annual food intake (%)	
Vegetable stuffs	5
Agave	2
Cactus fruit	1
Leguminous seeds and fruits	
Other small seeds	
Roots	
Other items	
Land animals	1
Rodents and reptiles	
Insects	
Deer and other large mammals	
Birds	
Marine animals	2
Shellfish	1
Fish	
Sea mammals	
Turtles	
Sea birds	
Bird and turtle eggs	
Typical food intake for the spring months of scarcity (%)	
Agave	4
Roots	
Other vegetable items (including second harvest and any stored seeds)	
Land animals	
Marine animals	4

All forms of the *A. shawii* complex do not dis play this habit, but it is characteristic of th group and is discussed more individually und the following accounts of the subspecies. In ac dition to customary morphological criteria, have taken both growth habits and breeding hal its into consideration in forming concepts of sp cies.

INDIGENOUS USES OF AGAVES

Indians in the tribes of California were a hunters and gatherers of wild plant foods. Th edible agaves were a very important resourc for them. Aschmann (1959) calculated th agaves comprised 28% of the food consume annually, but rose to 45% during spring month when other vegetable foods were scarce (Tabl 1).

Aschmann obtained his information from a chaeological reports and from the Hispanic a counts of the colonial missionaries. Probably th best of these accounts was written by Miguel d Barco, a pioneer Jesuit priest stationed at th Misión San Javier in the Sierra de la Gigant

DESERTICOLAE

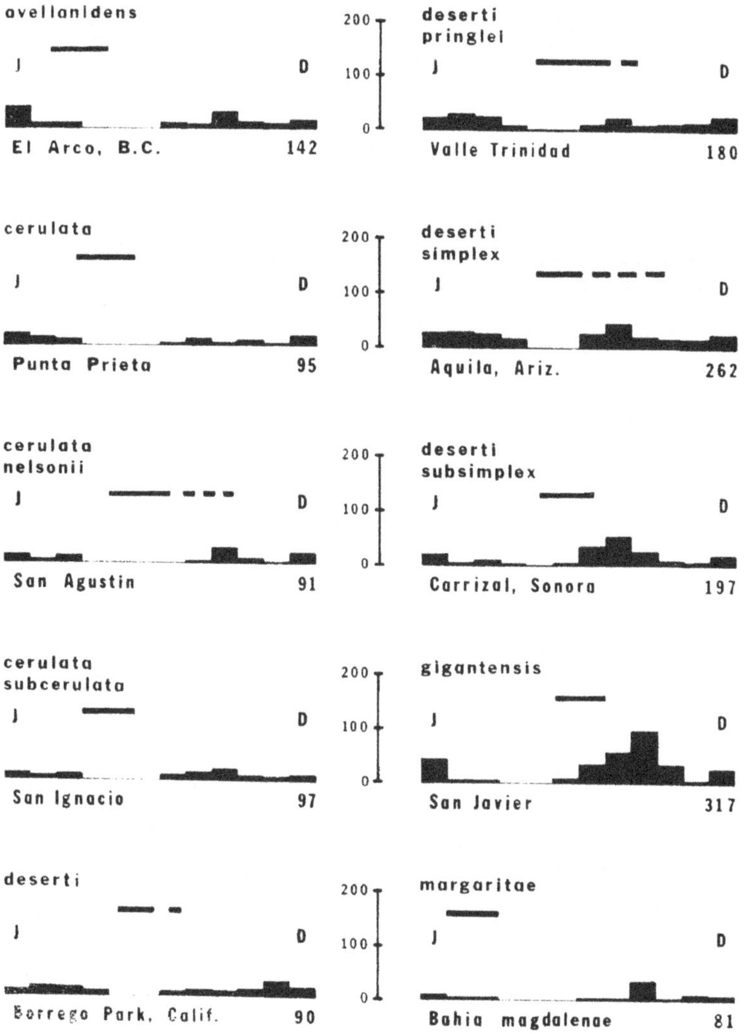

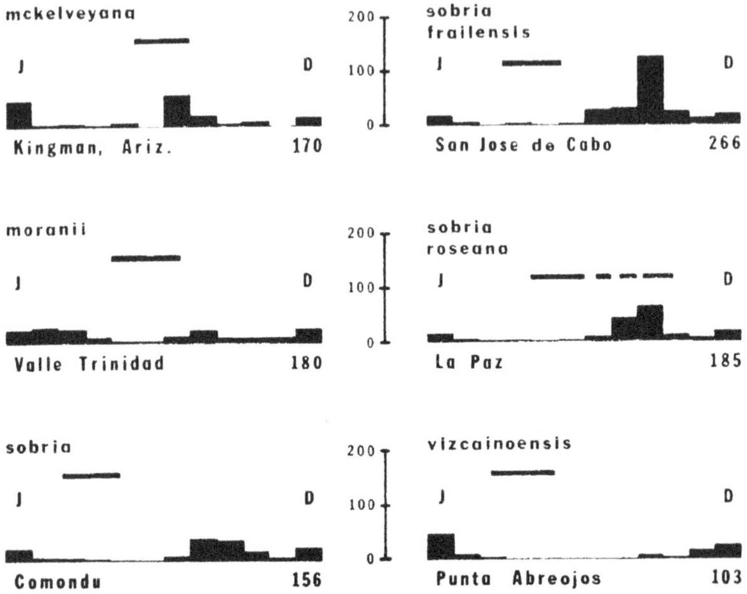

FIGURE 2. Rainfall (silhouettes) and flowering (bars) perimeters of the Deserticolae. Relevant meteorological stations with average annual rainfall given in millimeters. Data from *Atlas Meteorológico de México* (Servicio Meteorológico Mexicana 1939), Hastings (1964), Hastings and Humphrey (1969), and U.S. Weather Bureau. Flowering periods based on herbarium specimens and field observations, supplemented by plants in cultivation. Uncertainty expressed by broken lines.

from 1738 to 1768. After his return to Spain in the latter year, he wrote a lengthy account on the natural history of ancient California (Barco 1973). Other historians, such as Clavigero (1789), made use of Barco's report. Barco's descriptions of the uses of agaves by the native peoples, especially the Cochimi Indians inhabiting the Sierra de la Giganta region, is one of the most accurate and revealing I have read. Because of its excellence, I have translated from the Spanish to quote much of it here.

> The Californians did not use, nor do they now use the mezcal for making beverage. They employ it only for eating when the plant is ready or mature. They know when it has reached this stage, as the bud, which had its leaves tightly folded together, has spread out with its leaves open, separated one from another. In this stage, the last leaves to open are much smaller than the rest, and there is ready to sprout in place of the leaf bud, a shoot which, if the mezcal is not cut, will come out as thick as a man's arm, and which will rise to four to six *varas*. At its tallest it forms a great panicle composed of many small branches of yellow flowers which produce its seeds. When ripened, the whole plant dries up and the mezcal is worthless. The mezcal is not eaten raw but only roasted, and the whole operation, which is carried out by women, is done in the following way.
>
> Three or four women or more leave their rancheria or settlement in the morning, each provided with a carrying net on the shoulder over which it is maintained by some coarse cords which pass to the forehead of the woman. In this net, they carry the mezcales as well as whatever else needs to be carried all year. This net the Cochimi women call *uani*, and the Spaniards call it *aparejo*, because it serves to support and carry cargo. Over the forehead they put a piece of deer hide, doubled so that the cords do not cut into their foreheads when they are loaded. On such occasions they go bent over, lowering not only the head, but also inclining the shoulders and back because this reinforces the head against the weight of the cargo which it must resist. Besides the net, each woman carries a large heavy knife. In place of a machete to cut the trunk of mezcal, they have a little slab of hardwood, three to four fingers wide and two or three palmos [51–64 cm] long, in the form of a shovel blade but without point, but they bevel the end of it to cut the mezcal.

FIGURE 3. *Agave shawii goldmaniana* showing the old reclining trunks radiating out from a common mother rosette, long gone.

Arriving at whatever locality they are going, they separate and each one goes to look for mezcales. Finding a mature one, they take in hand their hardwood tool and apply the beveled end to the stem of the plant. The upper part of the plant having the leaves is the thickest, juicy and tender; this they call the head of mezcal and is what is eaten. Applying then the beveled end of the hardwood tool to the base of the head (where the green leaves begin), they strike heavily the upper end of the tool with a rock. Little by little they behead the mezcal. Then with their knives they cut off the leaves near the head of the plant, leaving a small basal portion of the leaves, about two fingers, which here are about a finger and a half thick, stuck to the head. When the head is so trimmed, it remains the size of that of a man. Then they look for more, and in the afternoon, each woman returns with eight or nine mezcales carried in the *uani*, which is a good load! Sometimes they travel one or two leagues [about 4–9 km], adding on top some wood to roast them.

Near the rancheria, they make fire and add to it some stones that are not too large. When the wood is consumed and the stones as hot as red coals, they spread the coals and stones a little with poles, and then, one by one, place the mezcales in accommodation onto the fire according to their judgement. Altogether it forms a mound, and they cover it with nearby hot earth, which reconcentrates the heat and delays its dissipation. Thus, it is left for twenty-four hours and, more frequently, for two nights and one day; and they dig it out all well cooked. The Mexicans and also all the Spaniards of New Spain call this method of roasting *tatemar*, and the thing that is roasted they call *tatema*. They use it to roast various things, principally heads of sheep and cows, etc., ... The meat remains tender, juicy, mild, and gustatory.

When the mezcales are taken out and left to cool, the woman has food for her family for three days, more or less, according to the number of persons. The bases of the leaves (the white growing tissue) which were left on the stems are also eaten, but we could better say they are chewed, as they have little more than a sweet juice and tough fibers which are not eaten. They spit out this bagasse like "tacos de escopete" (shotgun wads). This bagasse is not always lost, because the old men and women (who, although given food, are always hungry) recover these tacos that were ejected on the ground. When they are well dried, they ground them to powder between stones and eat them.

The remainder of the mezcal, freed of its leaves, is better esteemed as solid food. It is cut with a knife into slices and eaten with gusto; it is almost as sweet as conserves made with syrup. However, it leaves the mouth asperous, as with those of delicate palate who are not accustomed to it. In some regions, much finer mezcales are found that contain little or no fiber.

This is the most common food from October inclusive until April. In May they cease to eat it because the time of ripening is past or the new shoots are past and dry, and they will not start to ripen until the beginning of autumn. Then also they have no appetite for a hot food like mezcal in the warm season.

When these plants are not cut, they produce a flowering shoot, as we stated above, and in its flowers a good quantity of nectar is contained in the calyx of each flower. In order to collect this, they cut each small branch of flowers separately, turning them upside down over an *adda*. Each flower is emptied of its contents, and in a short while, the *adda* or wooden bowl is filled. This nectar is sweet but rather nauseous, but when boiled and the spume skimmed off, it becomes much better.

For the rest of the account, there are other species of mezcales that are not eaten, some because they are bitter and others because they cause bellyache. Finally, from the leaves of mezcales they extract fiber, with which they make both fine and coarse cords for making their nets and for other uses. . . . The roots of the mezcal are good for nothing, at least in California. [Barco 1973: 122 (transl. from Spanish)]

Additional notes on the indigenous uses of agaves are given below under appropriate group and species headings.

AGAVE MEASUREMENTS AND FLORAL IDEOGRAPHS

Measurements

Measurements of organs and parts have been widely employed in the classification of both plants and animals. Measurements provide basic data for direct and analytical comparisons between species and other taxonomic categories or comparison of individual specimens. Total length of leaf is measured from leaf axil to tip of spine. Width is generally given at its widest in mid-leaf, as in dried, pressed specimens. Teeth measurements include the broad corneous base and are only approximate because most teeth are curved or flexed in various ways, making precise measurements impractical. Height and width of rosettes, as well as height of panicle (overall length of inflorescence from ground level to tip), are also usually only approximate or estimates. Unusual variations are not given in descriptions. The ones given should be regarded as a general mean or average prevailing in the subject population or species.

Figure 4 shows the flower organs selected for measurement and how the measurements are made. However, my experience has shown that to be reliably comparable, the measurements must be made in a standardized way subject to certain qualifying conditions imposed by the developing flower. The flowers used should be normal, healthy, and precisely in the stage of anther dehiscence.

The flower bud grows slowly but opens and expands very rapidly during anthesis (expansion of the flower) within one day. Therefore, measurements for comparative purposes must be taken at about the time of anther dehiscence, which occurs during a few daylight hours. However, flowering begins at the base of the inflorescence and continues to move upwards in the inflorescence for two to four weeks, depending upon individual plants and species. One can usually find a series of flowers in late-bud stage, at anther dehiscence, and after dehiscence, all in one flower cluster. I prefer to have measurements of flowers taken just before, during, and right after stamen dehiscence so that maturation is bracketed. For this purpose, I collect and immerse the flowers in a pickling solution of one part formaldehyde, three parts 95 percent alcohol, and six parts water. Pint-size canning jars in cartons of one dozen are suitable for field work and accommodate all except a few of the very largest flowers. For permanent storage, they must be transferred to laboratory jars with noncorrosive resinoid tops. Labels are put inside the jars.

The pistil continues to grow after anther dehiscence, usually for two or three days,

TABLE 2. REPRESENTATIVE FLOWER MEASUREMENTS (IN MM) OF THE GENUS *Agave*, GROUP DESERTICOLAE. (*—measurements from dried flowers, relaxed by boiling; **—measurements from fresh or pickled flowers.)

Taxon & locality	Ovary	Tube	Tepal	Fil. ins. & length		Anther	Total length	Coll. No.
cerulata cerulata								
Catavinya	24	4.5 × 12	16 × 6	4–4.5	30	14	45**	23159
Laguna Chapala	31	6 × 12	20 × 6	6	32	19	57**	19973
San Luis Gonzaga Bay	26	3 × 13	17 × 7	3–4	35	16	45**	23298
Punta Prieta Valley	24	4 × 13	18 × 8	5	42	18	45**	23302
cerulata nelsonii								
S of San Miguel	35	3 × 12	17 × 4	3	40	17	55**	23324
S of San Miguel	27	5 × 13	21 × 7	5	30		54**	11179
cerulata subcerulata								
W of San Ignacio	28	2.5 × 12	18 × 5	3–2	34	18	48**	23170
E of San Ignacio	24	5 × 13	20 × 6.5	5	36	20	48**	23175
Isla San Marcos	27	3 × 12	19	3	33	20	50**	11892
deserti deserti								
Pinyon Flats	33	4 × 12	17 × 6	5	36	17	53**	19759
Pinyon Flats	31	3 × 11	15 × 4	3	37	16	50**	23326
San Felipe, Calif	26	4 × 15	17 × 8	6	28	15	48**	19940
San Felipe, Baja Calif.	38	4.5 × 12	18 × 7	4.5	40	18	60**	23286
deserti pringlei								
San Matias Pass	24	6 × 12	14 × 5	6.5	20	14	43**	19959
San Matias Pass	22	7 × 15	16 × 7	7	21	15	44**	19959
deserti simplex								
Harquahala Mt.	25	6 × 13	20 × 3–4	5–6	38	21	50**	23404
Harquahala Mt.	26	7 × 12.5	17 × 4.5	5–6	40	20	50**	23404
Providence Mt.	28	8	15	5–6		17	52**	3231A
mckelveyana								
Black Mountains	17	3 × 8	12 × 3	2–3	28	12	30**	21979
Hualapai Mt.	21	4.5–5	15 × 5	4	25	16	40**	22312
S of Wikieup	21	4 × 12	12 × 4	2–3	29	12	36**	23000
subsimplex								
Libertad	23	5 × 11	17 × 8	4–5	30	16	45**	3880
Desemboque	25	3 × 10	14 × 7	2	28	15	43**	14171
sobria sobria								
Comondú	25	3 × 9	21 × 3	3	47	22	48**	1989
Comondú	29	4 × 10	17 × 3	3	35	20	49**	11882
sobria roseana								
Espíritu Santo	40	5 × 12	22 × 5	4–5	42	23	66**	11277
La Paz	26	3.5 × 11	18 × 5	3.5	29	19	46**	11869
sobria frailensis								
Punta Frailes	37	3.5 × 12	19 × 4	3	40	20	58**	11264
Punta Frailes	31	2–3 × 12	24 × 6	2–3	38	23	57**	11257
Punta Frailes	24	4 × 11	17 × 4	4	32	18	43**	11858
moranii								
Bajada of Sierra San Pedro Mártir	41	6 × 13	23 × 7	6–5	46	21	70**	23287
Bajada of Sierra San Pedro Mártir	40	5 × 13	24 × 6	5	43	21	68**	23287
Bajada of Sierra San Pedro Mártir	26	4 × 13	19 × 7	4.5	32	17	49**	23287

Table 2. Continued.

Taxon & locality	Ovary	Tube	Tepal	Fil. ins. & length		Anther	Total length	Coll. No.
avellanidens								
Paraíso	31	6 × 15	16 × 4.5	6	38	18	52**	23187
Mesquitál	34	5 × 14	19 × 6	5	46	18	57**	23186
Mesquitál	18	4 × 12	15 × 5	4	35	15	37**	11932
Calmallí	35	6 × 15	22 × 7	6	45	23	65**	23184
gigantensis								
Sierra Palmas	30	5 × 13	22 × 5	5	44	21	56**	23320
Sierra Palmas	25	4 × 12	23 × 5	4	25	24	51**	23320
Sierra Palmas	27	5 × 12	18 × 5	5	37	20	50**	10324
vizcainoensis								
Cerro Tordillo	39	13 × 15	21 × 4	9 & 8	50	21	72*	7469
Picachos Santa Clara	35	8 × 15	26 × 5	6	63	25	68*	7713

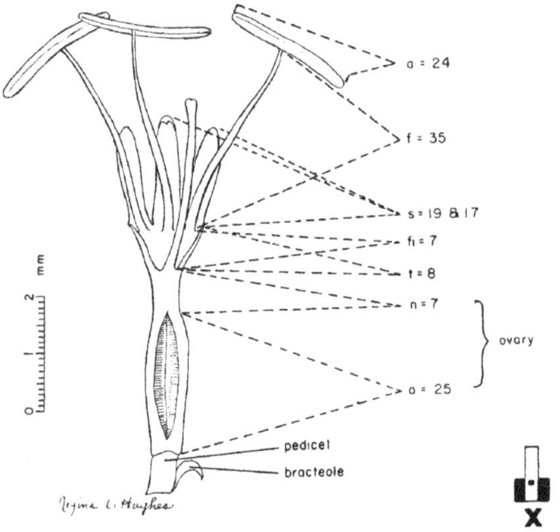

o - ovary body length, n - neck of ovary length, t - tube length, fi - filament insertion (measured to bottom of t - tube), s - tepal lengths, f - filament length, a - anther length, tl - total length

Figure 4. Cross section of agave flower with parts measured and a tube/tepal ideogram, *x*. The white column represents the tepal, the black the tube, and the black square the insertion of the filament in the tube.

when the stigmas open and are receptive. Because of this later growth, measurements of pistils at anther dehiscence have no comparative value and have not been made.

Floral Ideograph

The measurements can be analytically applied to form the symbolic floral ideograph. The various forms of ideographs and their applications were very ably described and applied by the late Edgar Anderson (1949). The form ideographs may take is more or less determined by the organs which the taxonomist finds best representative of the variability that expresses speciation, introgression, or developmental features of a related group of organisms. The ideograph here selected shows the tube/tepal ratio, together with the insertion of the filament within or on the rim of the tube and, more rarely, on the base of the tepal. The ideograph is a kind of taxonomic shorthand that throws into perspective the selected characters, so strikingly apparent in Figures 5 and 6.

Old dried flowers are shrunken and distorted by the growing ovary, and measurements of them are unreliable. They are little improved by relaxing. Berger, at La Mortola Botanical Garden, pressed split flowers under great pressure and, so treated, their proportions remained fairly reliable. So are split flowers from the pickling jar.

The junction of tube and tepals is a particularly critical level, as the length of both organs derive from this. In measuring the length or depth of tube, one must note the various tepal sinuses because some may be deeper than others in the same flower. The upper limit of tube is taken at the median level of the sinuses. Where there is marked overlap of tepal bases in the sinuses, the median point of the overlap is taken as the upper limit of tube. There is, thus, a built-in indefiniteness about many tubes that may amount to a 1-mm variance or a margin of about 10 percent inexactness for tube measurement. Also, one must watch for damage by insects and birds, which sometimes split the sinuses. With measurements of three or four flowers, made after splitting the flower lengthwise, these difficulties are largely overcome, and the mean or median measurement is suitable for comparisons with other collections.

Diameter of tube is its widest extent, usually at about the level of filament insertion.

Width of outer tepal is the median width just above the basal widening, or at about ⅓ to ½ tepal length above the tepal sinuses. This is bound to be a general-impression kind of measurement as it depends on degree of wilting or degree of inrolling and cannot, therefore, always be precise.

Where there is a marked difference in length of outer and inner tepals, the length of the outer is given first. Width of the inner tepal is not given, but its form, shape, or other qualifying characteristics are indicated in the description of species.

Floral ideographs and the measurements on which they are based are given for each group, following the group description.

GENUS AGAVE

Agave LINNAEUS, Species Plantarum 1:323. 1753.

Succulent rosettes, monocarpic or polycarpic, perennials or multiannuals with long-lived leaves, frequently suckering at base and occasionally bulbiferous in the inflorescence; roots hard fibrous, radiately and shallowly deployed; stems thick, very short, usually shorter than the terminal bud, simple or branched; leaves large, generally succulent, spine-tipped, margin armed or unarmed with teeth; inflorescence tall, bracteate, scapose, spicate, racemose, or paniculate with flowers in umbellate clusters; flowers mostly large, generally proterandrous; perianth tubular to shallowly funnelform, the six segments erect to variously curved, similar or dimorphic, imbricate in the bud; stamens six, exserted; filaments long, inserted in tube or on tepal bases; anthers versatile; ovary inferior, three-celled, succulent, thick-walled with numerous axile ovules in two rows; pistil elongate, filiform, tubular; stigma three-lobate, papillate glandular; fruit a dehiscent, loculicidal capsule; seeds flattened, black.

Subgenus *Agave*: inflorescence paniculate, flowers in umbellate clusters on lateral branches. Type species of genus and subgenus: *Agave americana* L. Sp. Pl. 1:323. 1753, described from cultivated European plants.

KEY TO THE GROUPS

1a. Leaves ensiform, 10–20 times longer than wide at mid-blade; rosettes suckering widely with elongate rhizomes .... Datyliones (p. 97)

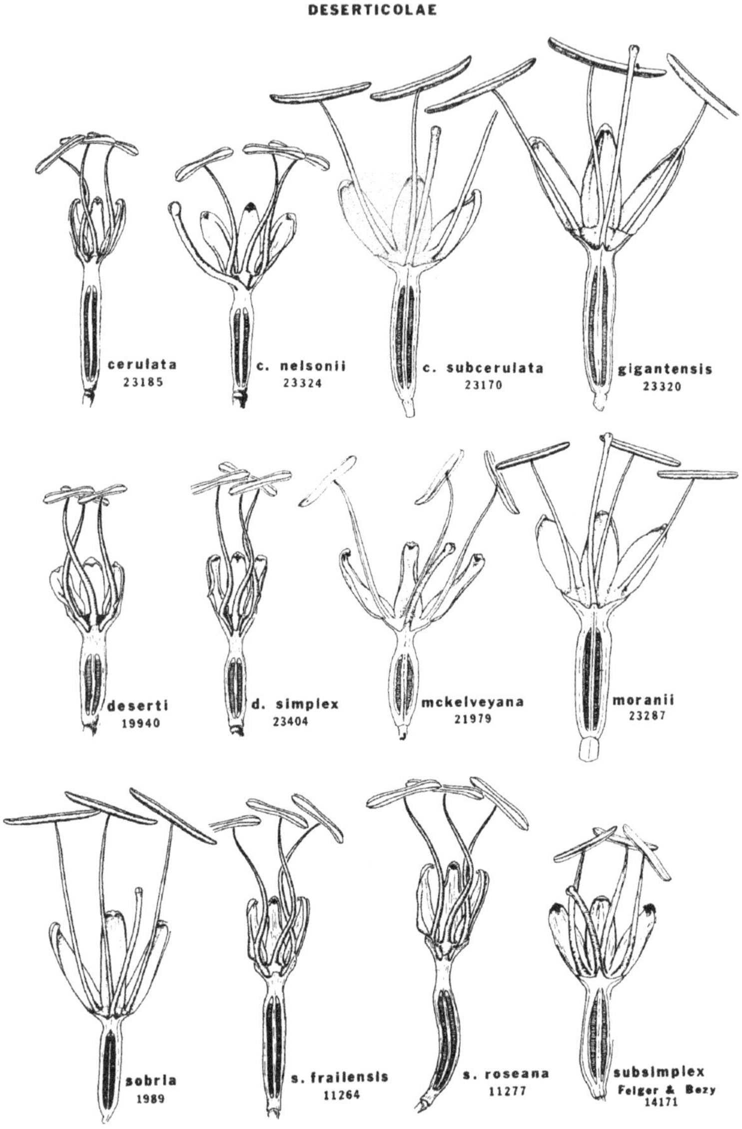

FIGURE 5. Flower sections representative of the Deserticolae.

DESERTICOLAE

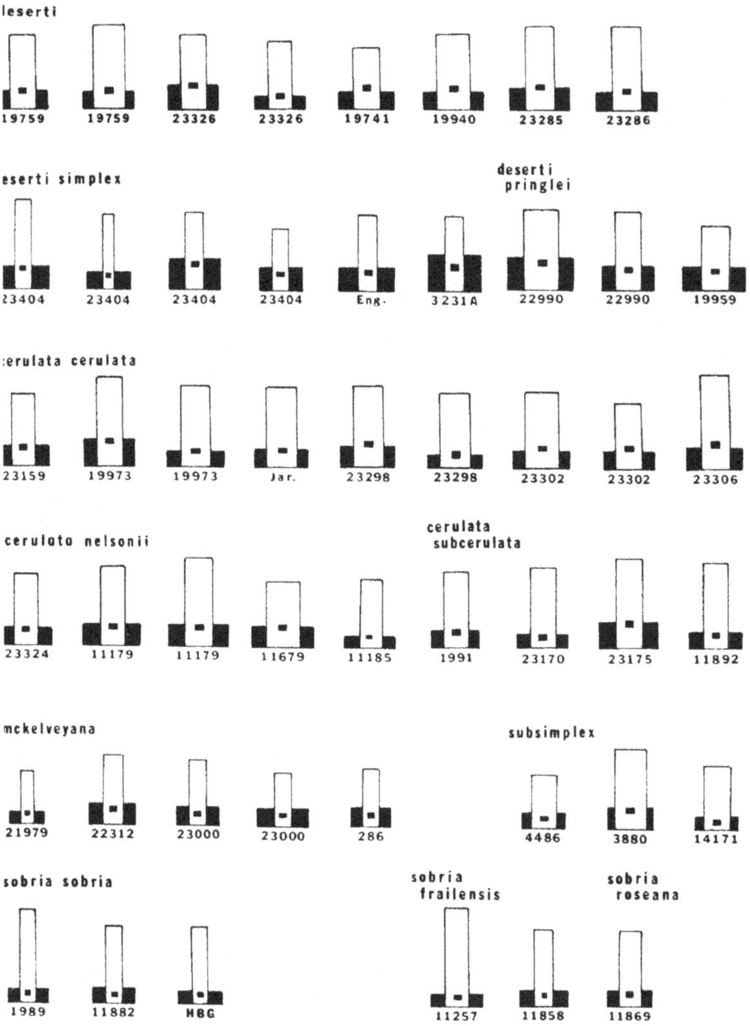

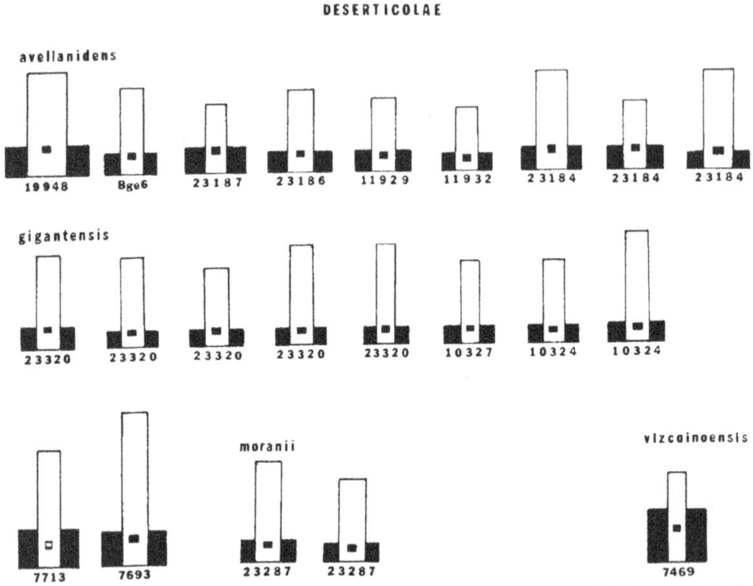

FIGURE 6. Floral ideographs of the group Deserticolae, showing relative proportions of the tube (black) to outer tepal (white column), and level of insertion of filament (black square). Measurements are listed in Table 2.

1b. Leaves not ensiform, only 2–10 times longer than wide; rosettes not suckering or bearing suckers appressed to the base 2

2a. Stems short, usually about as wide (leaf axil to leaf axil) as long (except *A. capensis*); branches of the panicle not borne in large succulent bracts, the umbels small- to medium-sized; flowers smaller (40–70 mm long), tubes shallow or broad and open 3

2b. Stems elongate, longer than broad; branches of the panicle usually borne in large succulent bracts, the umbels broad and massive; flowers large (70–100 mm long), tubes deep .... Umbelliflorae (p. 82)

3a. Plants generally green, nonsurculose or branching from leaf axils; leaves relatively smooth; flowers red to purplish in bud, opening to light orange, campanulate with broad open tubes and curved tepals .... Campaniflorae (p. 70)

3b. Plants light gray to yellowish green, surculose or single, not branching from leaf axils; flowers greenish yellow, funnelform with short tube and ascending tepals .... Deserticolae (p. 13)

GROUP DESERTICOLAE

TRELEASE, Missouri Bot. Gard. Rep. 22:45. 1912.

Plants small- to medium-sized, glaucous gray to greenish, freely suckering, or medium-sized to large, green, nonsurculose, the rosettes acaulescent to short caulescent; leaves rigid, coarsely fibered, with thick cuticle, narrowly lanceolate and with weak, easily detached teeth, or broader and with firmer teeth; panicle narrow with short lateral branches, dry scarious peduncular bracts, and small umbellate flower clusters; flowers small with very short open tube, the tepals about equal and three to five times as long as the tube; spring flowering; capsules small to medium, freely seeding. Sonoran Desert region in southeastern California, Arizona, Baja California, and Sonora.

The leaf surface in the Deserticolae is generally light colored, glaucous gray or yellowish or light green, minutely sculptured, the cuticle rather thick and rough with variable papillae. The stomata are relatively dense, ranging from 30 to 50 per mm² on the upper leaf surface. The stomata are generally depressed and frequently so close together that the stomatal depressions are confluent, forming grooves in the cuticle, transverse to the long axis of the leaf, as in *A. deserti* and *A. cerulata*, or aligned with the leaf axis as in *A. sobria sobria*. The margin of the stomatal pore is generally irregular and is without the stomatal rim. However, among the seventeen taxa of this group, there are many exceptions to the above generalities. The intrasectional variability is reflected in the epidermal morphology even as it is in the gross morphology of this complex group. For further particulars on epidermal morphology, the reader is referred to the companion study (Gentry and Sauck 1978).

The Deserticolae do not display close morphological relation to any other *Agave* group. Certain other taxa, such as *A. neomexicana* and *A. gracilipes* of the Parryanae, are also small-leaved xerophytes, and the latter has an ideograph reflecting its short tube, much like the Deserticolae. However, the geographic position and other morphological characters of *A. gracilipes* recommend its inclusion in the Parryanae. The Marginatae are also similar to the Deserticolae with their very short tubes and relatively long tepals, but they belong in the subgenus *Littaea* and have other important distinguishing characters, such as their marginate leaves and filament-clasping tepals, while their geographic distribution is distinct.

Perhaps the closest relatives of the Deserticolae are the anomalous *A. utahensis* group, occupying an adjacent area. Similar characters in the two groups are the short flower tubes, the short branches of the panicle, the weakly attached teeth, and the prolific stooling rosettes. However, other characters of flower, such as its connivent form, and of fruit and leaves demonstrate a wide divergence between the groups. The common group ancestor, assumed to have existed on general biological theory, was remote in biologic time and in physiographic if not geographic space.

Agave deserti deserti occupies a centric position in Deserticolae, certainly in physiographic position, and interpretively in morphological relations. Several of the taxa, such as *A. pringlei* and *A. deserti*, are too close to separate specifically, while others, such as *A. sobria* and *A. moranii*, are quite distinct. The latter, along with *A. avellanidens* and *A. gigantensis*, form a distinctive group to themselves, again based mainly on rosette and habit characters. They could well be treated as a subsection, as Berger 1915 and most other German taxonomists would probably have done upon cognizance of the groups. However, I prefer to keep the groups simple wherever possible. The Deserticolae forms a natural phylogenic group on both morphological and historical grounds.

Flower structure is quite stable or uniform throughout the Deserticolae (Figs. 5, 6). The differences or variations in the floral organs are not always critical for defining species, as I had hoped before the respective series of flowering specimens were assembled. Variation in flower structure was sometimes found to be as great within local populations as it was between populations judged to be separate species on other criteria. This is the case, for instance, with *A. deserti* and *A. cerulata*. These two complexes are difficult to separate by morphological characters, although respective populations appear distinct when viewed overall in the field. The chemical compositions of sapogenins appear to substantiate their specific separation (Tables 3 & 4).

However, there are some cases where flower size, proportions, and shapes were correlative with vegetative characters and were useful for separating species, as with the small flowers and narrow tepals of *A. sobria* and *A. mckelveyana*. Leaf form, size, habit, and other variation patterns, combined with geographic distribution, are the principal criteria used in separating the taxa.

Key to the Species of the Deserticolae

1a. Plants surculose; leaves small, usually narrow, less than 10 cm wide, light glaucous gray to yellowish; teeth frequently fragile; panicles short, mostly with only 8–15 umbels 2

1b. Plants not surculose; leaves broad, 10–25 cm wide, green to glaucous gray;

teeth firmly attached; panicles elongate with 20–40 lateral umbels 8

2a. Plants of the northern sonoran Desert; leaves generally gray to light green, broader, less acuminate 3

2b. Plants of peninsular Baja California; leaves generally yellowish to gray, narrower, more acuminate; teeth with a brown ring around base, or leaves larger, frequently cross-zoned 5

3a. Leaves larger, mostly 25–40 × 5–10 cm; teeth very feebly attached; spine decurrent for several cm and confluent with uppermost teeth; flowers 40–60 mm long; filaments inserted on base of tepals or in orifice of tube .... *deserti* (p. 15)

3b. Leaves smaller, 3–5 cm broad at mid-blade; teeth feebly or firmly attached; spine very shortly decurrent and not confluent with uppermost teeth; flowers 30–50 mm long; filaments inserted in orifice of tube 4

4a. Leaves 18–30 × 3–5 cm; teeth firmly attached; flowers 30–40 mm long; flowers yellow; tepals 3–4 mm broad in middle, conduplicate, strongly cucullate; ovary constricted below tube. Northwestern Arizona .... *mckelveyana* (p. 25)

4b. Leaves 15–25 × 3–6 cm; teeth weakly attached; flowers 40–50 mm long; flowers partly pink or red; tepals 6–8 mm broad in middle, plane or rounded, slightly cucullate; ovary not constricted below tube. Coastal Sonora .... *subsimplex* (p. 27)

5a. Leaves small, mostly narrow, long-acuminate, 3–6 cm broad (rarely 7–8 cm); yellowish to light glaucous gray; teeth relatively small, mostly 3–5 mm long, very weakly attached and ringed by a brown margin below base. Mid-peninsular region .... *cerulata* (p. 35)

5b. Leaves larger, or at least broader, ovate to linear-lanceolate, mostly 5–12 cm broad at mid-blade, green to gray glaucous; teeth usually large, 8–20 mm, variously flexed, firmly attached, not ringed by a brown basal margin 6

6a. Flower tube shallow, 3–5 mm deep; leaves mostly 40–60 cm long, sometimes conspicuously cross-zoned. Sierra de la Giganta region .... *sobria* (p. 48)

6b. Flower tube 8–12 mm deep; leaves short, usually less than 50 cm long, not cross-zoned. Outer coastal region 7

7a. Leaves short, less than 25 cm long, ovate to oblanceolate, the margin prominently mammillate and with long (8–15 mm) slender teeth .... *margaritae* (p. 48)

7b. Leaves 25–40 cm long, lanceolate, the margin nearly straight to undulate, with shorter, more flattened teeth .... *vizcainoensis* (p. 67)

8a. Leaves 70–120 cm long, triangular long-lanceolate, deeply guttered, the margin nearly straight; terminal spine subulate, light gray; peduncle sometimes swollen below the panicle. South and east slopes of the Sierra San Pedro Mártir .... *moranii* (p. 58

8b. Leaves 40–70 cm long, broadly lanceolate, plane to concave above, margin lightly to deeply sinuate; spine coarser, dark brown to dark gray, frequently sinuous; peduncle not swollen below panicle 9

9a. Leaves green, broadly linear-lanceolate, not or scarcely narrowed towards the base; panicle generally occupying one-half the shaft with 25–35 laterals; flowers yellow to orange. Central peninsula .... *avellanidens* (p. 61

9b. Leaves light green to glaucous gray, ovate acuminate to spatulate, conspicuously narrowed toward the base; panicle in upper one-third of shaft with 18–25 laterals; flower buds whitish or pale green, opening pale yellow. Southern peninsula .... *gigantensis* (p. 63

Agave deserti

Agave deserti Engelm. ssp. **deserti**

(Figures 7, 8, 9 (*upper*); Tables 2, 3)

Agave deserti Engelm. Trans. Acad. Sci. St. Louis 3:31(370. 1875.

Agave consociata Trel. Missouri Bot. Gard. Rep. 22:5: 1912.

Medium-sized, light gray, sparingly or prolif ically suckering rosettes mostly 30–50 cm tall 40–60 cm in diam.; leaves variable, mostly 25 40 × 6–8 cm, lanceolate to linear-lanceolate scarcely narrowed above the broad claspin base, acuminate, gray glaucous to bluish glau cous, often cross-zoned, thick, rigid, concav

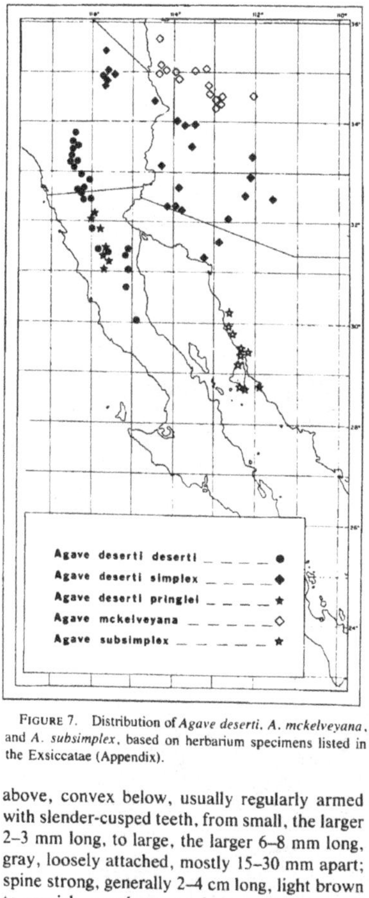

FIGURE 7. Distribution of *Agave deserti*, *A. mckelveyana*, and *A. subsimplex*, based on herbarium specimens listed in the Exsiccatae (Appendix).

above, convex below, usually regularly armed with slender-cusped teeth, from small, the larger 2–3 mm long, to large, the larger 6–8 mm long, gray, loosely attached, mostly 15–30 mm apart; spine strong, generally 2–4 cm long, light brown to grayish, openly grooved above, decurrent to the first or second tooth above; panicles generally 2.5–4 m tall, on slender shafts with scarious triangular bracts 8–15 cm long and 6 to 15 short laterals with small umbels in upper 1/4 to 1/5 of shaft; flowers yellow, 40–60 mm long; ovary 22–40 mm long with a slightly narrowed neck 4–6 mm long, greenish; tube shallow, spreading, 4–6 mm deep, 12–15 mm wide, lined with a thick nectiferous disk; tepals equal, 14–20 mm long, 6–8 mm wide, light yellow, broadly linear, spreading at anthesis, rounded and abruptly hooked inward at apex; filaments 25–35 mm long, inserted at base of tepals; anthers 13–18 mm long, yellow; capsules ovoid to oblong or obovoid, mostly 3.5–5 cm long, 1.5–1.8 cm wide, thick walled, short-stipitate; seeds black, 4 × 5 m.

Type.—Based on Emory in 1846 and Hitchcock and Palmer in 1875, collected on Rancho San Felipe, San Diego County, California.

Distribution, Variability, and Relationships.—*Agave deserti* is a large variable complex, the limits of which are hard to define. The geographic distribution is mapped in Figure 7 and documented in the Exsiccatae. While the extensive population of *A. deserti* at the type locality along San Felipe Creek (Fig. 8) is morphologically homogenous, there are other localities showing wide variability in leaf form. One of these localities is Pinyon Flats by the northwest slopes of Santa Rosa Mountain along Route 74 in southern California.

Some of the variability appears attributable to hybridization, such as those populations on Pinyon Flats and in the San Matias Pass in Baja California. The population from the latter area is discussed below under *A. deserti pringlei*. Variability in the Pinyon Flats population appears to be a relic condition of geographically past introgression between two disparate genotypes, which are not recognizable today. Cave (1964:166) has reported a collection (*Hutchinson 710*) from Baja California with chromosomes $n = 59$. This is a polyploid condition, as the basic number in *Agave* is $n = 30$. The genetic variability in the *Agave* complex indicates a high potential for species evolution, as examples are the populations of *A. deserti pringlei* and *A. cerulata nelsonii* in our present Recent Period.

Ivan Johnston (1924) included several of Trelease's taxa as synonyms under *A. deserti*, viz. *A. pringlei*, *A. nelsonii*, *A. consociata*, and *A. dentiens*, thus forming a kind of superspecies category. After considerable study and vacillation, I am using Trelease's name, *A. pringlei*, as subspecies to designate scattered populations bordering the main complex on the south and west of the *deserti* area. Another group in and adjacent to Arizona is segregated as subspecies *simplex*, the name referring to its more simple habit.

A. cerulata Trel. is closely related, but ap-

FIGURE 8. *Agave deserti* near the type locality along the San Felipe arroyo, San Diego County, California. (*Upper*) The desert habitat; (*lower*) detail of drought-shrunken leaves. Otherwise, these are robust plants in deep sandy soil.

TABLE 3. SAPOGENIN CONTENT IN *Agave deserti*. Collated from Wall (1954) and Wall et al. (1954a, 1954b, 1955, 1957). Sapogenins are given in percentages on dry-weight basis: hec. = hecogenin; git. = gitogenin; man. = manogenin; tig. = tigogenin; w.pl. = whole plant; fl. = flower; dead = leaf of dead plant.

				Sapogenins				
Coll. no.	Source locality	Month coll.	Plant part	% total	% hec.	% git.	% man.	% tig.
	Agave deserti deserti							
	CALIFORNIA							
10041	San Felipe Narrows	Dec.	leaf	0				
10041	San Felipe Narrows	Dec.	stem	0				
10044	San Felipe Narrows	Dec.	w.pl.	0.1	50		50	
10051	Pinyon Flats	Jan.	fl.	0				
10051	Pinyon Flats	Jan.	fruit	0.6	X			
10051	Pinyon Flats	Jan.	seed	0				
10051	Pinyon Flats	Jan.	fruit	0.6	70	5	5	20
11650	Pinyon Flats	Apr.	leaf	0				
11652	Pinyon Flats	Apr.	leaf	0.75	20	14		66
12436	Pinyon Flats	Dec.	leaf (dead)	0.6	X	X	X	
	Agave deserti pringlei							
	BAJA CALIFORNIA							
10287	S of San Pedro Mártir	Mar.	leaf	0				
10287	S of San Pedro Mártir	Mar.	stem	0				
	Agave deserti simplex							
	ARIZONA							
9947	Harquahala Mt.	Nov.	leaf	tr.				
9947	Harquahala Mt.	Nov.	stem	0				

pears to be a separate specific complex, difficult to separate morphologically. Reasons and characters for separating it from the *deserti* complex are given under *A. cerulata* below. *A. cerulata* does not belong with *A. sobria* Bdge. where Johnston placed it. The following outline separates three main *A. deserti* geographic populations, none of which is in all individuals morphologically distinct.

Key to Subspecies of *Agave deserti*

1a. Leaves mostly 25–40 cm long, 4–7 times longer than broad, moderately acuminate, the margins usually straight; teeth weakly attached; spine decurrent as a corneous margin only to the upper first or second pair of teeth. Tube 3–8 mm deep 2

1b. Leaves mostly 40–70 cm long, 8–12 times longer than broad, long-acuminate, the margin straight; teeth firmly attached; spine conspicuously decurrent in a corneous margin frequently to the mid-blade or even below. Tube 5–8 mm deep. San Matias Pass and vicinity .... *pringlei*

2a. Rosettes copiously surculose, forming large clones; flower tube 3–5 mm deep; filaments inserted on base of tepals. Western side of the Gulf of California .... *deserti*

2b. Rosettes generally single, rarely with 1–3 offsets; flower tube 5–10 mm deep; filaments inserted in orifice of tube below tepal bases. NE of the Gulf of California .... *simplex*

Uses and Chemistry.—*Agave deserti* is among the more edible of the agaves. Castetter et al. (1938) wrote an excellent account of uses of agave by the Indians of the American Southwest. Their map of "mescal finds" (ibid.:37), including "mescal pits," shows that *A. deserti* was eaten and its fibers used throughout the *deserti* area. "Mescal pits" are very common and have been traced as far south as the latitude of Punta San Fermin in the Gulf of California (ibid.:60). This is in Baja California at the southern margin of the known distribution of *A. deserti* (e.g., *Gentry & McGill 23286* on the San

FIGURE 9. *(Upper) Agave deserti deserti* on the desert plain southeast of Cerro Borrego, Baja California. The high small panicles are characteristic in this region. *(Lower) A. deserti pringlei* in San Matias Pass with green and pale-glaucous forms growing side by side.

Pedro Mártir bajada). Barrows (1967:59) stated that the Coahuilan Indians of southern California made much use of *A. deserti* for food and fiber. They called the plant "a-mul," sections of the flowering stalk "u-a-sil," the leaves "ya-mil," and the yellow blossoms "amul-sal-em," all of which were cooked in various ways and eaten. Barrows regarded the agaves as one of the principal plant resources of southwestern Indian tribes.

Other tribes known to use *A. deserti*, as well as other species in their respective areas, are the Pimas and Papagos of both Arizona and Sonora, the Yumans, Kamias, Chemehuevis, and Yavapais along the Rio Colorado, and the Digueños and Cocopahs of northern Baja California. Other adjacent tribes, such as the Utes and Apaches, received cooked *A. deserti* in trade. The earliest missionaries in the 17th century reported the uses of agave in the *deserti* area, for example, Padre Kino (Bolton 1919). Other early reports, such as that of the German missionary Baegert (Rau 1864), who was stationed near the 25th parallel on the western slope of the Sierra de la Giganta in the 1770's, and Barco (1973), who was stationed at San Javier between 1738 and 1768, corroborate the widespread use of agaves in Baja California. The Indians of the peninsula were hunters and gatherers; they did not practice any agriculture, and the wild agaves were of major importance for their survival in this aridly inhospitable land.

The absence of mescal pits south of Sierra San Pedro Mártir does not mean that agaves were not eaten south of this latitude. The account of Miguel del Barco (1973:123), tells us why there are no mescal pits southward. It just was not the custom of the Cochimi to dig cooking pits. They roasted their mescal upon the ground "en forma de monton." The Cochimi ranged through a large central section of the peninsula and spoke a different language than the northern Digger Indians. The Cochimi guides on Link's expedition of 1766 (Burrus 1966) found they could not communicate with the San Pedro Mártir people. The southern end of this sierra marks a division margin between the two language-and-cultural groups; so also do the southern limits of mescal pits.

Agave deserti is important to wildlife; birds visit the flowering stalks for nectar and insects. Small rodents live about the plants and are protected by the armed rosettes. The pack rats, *Neotoma*, make nests among the clumps of rosettes, gnaw through the leaves, and eat the flowers and perhaps the seeds. Wild bighorn sheep, like cattle, eat the new-flowering shoots. As a wildlife resource growing in our most arid deserts, *Agave deserti* deserves conservation and perhaps should be established on desert mountains where it is now lacking.

Table 3 gives the sapogenin content found in *Agave deserti* by chemists of the Agricultural Research Service in Philadelphia, Pennsylvania (Wall 1954; Wall et al. 1954a, 1954b, 1955, 1957). Hecogenin is the leading steroid. The steroid content is erratic to absent in leaves, absent in the edible stems, and appears to run about 0.6 to 1% in the fruits.

Agave deserti Engelm. ssp. **pringlei** (Engelm. ex Baker) Gentry, stat. nov.

(Figures 7, 9 (*lower*), 10; Tables 2, 3)

Agave pringlei ENGELM. EX BAKER, Handbook Amarillid. 182. 1888.

Green or whitish, cespitose, rather strict rosettes, offsetting closely from root crown, 4–7 dm tall, 5–8 dm wide; leaves narrowly triangular lanceolate, very long acuminate, mostly 40–70 × 5–7 cm, thick and widest at base, deeply crescentic in cross-section, guttered and thinner above, green to yellowish green or light glaucous gray; spine 3–4 cm long, aciculate, reddish brown to light gray, usually narrowly grooved for short distance at base, decurrent as a corneous margin to upper teeth or to the mid-blade; teeth rather regularly spaced 1–2 cm apart, mostly 5–10 mm long, slender and little curved, more rarely smaller or more broadly flattened and irregularly flexed; panicle 3–6 m tall with 10–15 lateral branches, very narrow or more ample with longer laterals and larger umbels; flowers 40–60 mm long, yellow; ovary 20–35 mm long, fusiform, roundly angulate, with furrowed neck; tube ample, 5–8 mm deep, bulging, strongly furrowed from tepal sinuses; tepals nearly equal, 15–20 × 5–7 mm, spreading, linear, rounded over finely cuculate apex, the inner wider than outer and with broad low keel; stamens small, filaments inserted on rim of nectary in orifice of tube; capsules 3.5–5.5 × 1.2–1.5 cm, mostly oblong, beaked, short-stipitate; seeds as for species.

Type.—"Central mountains of Lower California, alt. 6000 ft., Orcutt! Described from a dried specimen sent to Kew by Mr. C. G. Pringle." Probably from the Sierra Juárez in 1882, when Orcutt collected there.

The type, perforce, is only nomenclatorial in scope. The type specimens are but a pitifully small expression of the peninsular highland populations. In the description above, I have drawn this taxon's outline from the listed specimens (see Exsiccatae) to include the apparent in-

FIGURE 10. *Agave deserti pringlei* in San Matias Pass, Baja California. (*Upper*) Detail of rosette and cross section of leaf bases where specimens were cut. (*Lower*) Rosette of a green clone in flower, April 1963.

trogressions with *A. deserti deserti* and with *A. moranii* as well. Without including some morphological characters of *A. moranii*, *A. deserti pringlei* would appear as a green, highland ecotype of *A. deserti*.

The variable population in and about San Matias Pass shows relation to *A. deserti deserti* in the tall, very narrow panicle and the shorter white glaucous leaf, which are present in differing combinations in some of the clones. Relation to *A. moranii* appears in the denser broader panicles and yellowish-green leaves of other clones, as well as leaf elongation in both green and whitish forms. Other specimens on the western slopes of the Juárez-Mártir mountain axis are very similar to *A. deserti deserti* in leaf size and form, but have atypically green leaves. The type collections of *A. pringlei* appear to be of this green form. The population in and about San Matias Pass appears to be a "hybrid swarm," and individual genotypes here, as well as elsewhere, reflect genetic factors of one or the other "parent" in varying degree and combinations. Subspecies *pringlei*, with its surculose habit, slender panicles, and smaller size is more closely related to *deserti* than it is to the larger, nonsurculose, and diffusely paniculate *moranii*. The deeper tube with its deeper insertion of filaments is perhaps the most unique character distinguishing *A. deserti pringlei* from its neighbors, but incidence of this character appears to be sporadic, judging from the limited flower specimens available.

Agave deserti ssp. **simplex** Gentry, ssp. nov.
(Figures 7, 11, 12; Tables 2, 3)

Small- to medium-sized, mostly single, light green to light glaucous gray, compact, short-stemmed rosettes; leaves mostly 25–40 × 6.5–10 cm, rarely to 50 cm, lanceolate, firm, rigid, mildly concave above, with undulate to crenate margins, broadest in mid-blade; spine stout, subulate, 3–4 cm long, broadly grooved above, dark brown to light gray, decurrent for several centimeters to upper second or third pair of teeth; teeth in mid-blade mostly 5–8 mm long, 1–3 cm apart, brown to pruinose gray, frequently brown and gray-ringed about base, variously curved or reflexed; panicles 4–6 m tall with 8 to 15 short umbellate branches in upper third of shaft, the peduncles stout, large reflexed bracts below rapidly decreasing in size upward and persisting erect, scarious; flower clusters small, congested, pale yellow to ferruginous in bud; flowers 40–60 mm long, yellow with pale-green ovary; ovary 20–30 mm long, fusiform, with constricted grooved neck 4–6 mm long; tube short, 5–8 mm long, open funnelform, grooved, angular; tepals subequal, 15–20 × 4–5 mm, erect to spreading, rather thin, linear, conduplicate wilting and incurved, apiculate-cucullate, the outer slightly longer, flat to rounded on back, the inner with broad low keel; filaments 30–42 mm long, unequally inserted below tepal bases high in tube, slender, elliptic in cross-section, pale yellow; anthers 15–21 mm long, yellow, regular, centric; capsules 3.5–4.2 × 1.5–2 cm, oblong to obovoid, stipitate, rounded to rostrate at apex, rather thin-walled; seeds 5–6 × 4–4.5 mm, lunate to lacrimiform, dull or shiny black, the margin with a raised, irregularly fluted wing.

Type.—*Gentry 23404*, N slope of Harquahala Mountain, 12 miles [ca. 19 km] W of Aguila, Yuma Co., Arizona, 12 June 1974; deposited in US, isotypes in DES, MEXU, ARIZ.

Planta parva vel media plerumque simplex tandem paucisurculosa; foliis plerumque 25–40 cm longis 6–10 cm latis ad medium raro ad 50 cm longis, linearibus vel lanceolatis glaucoviridibus rigentibus undulato-marginatis; spinis marginalibus ad medium foliorium plerumque 5–8 mm longis, 1–3 cm separatis, debilibus flexuosis acicularibus bruneis vel albicantibus; spina terminali 3–4 cm longa subulata castanea vel grisea supra late canaliculata; inflorescentia paniculata angusta 4–6 m alta cum scapo, ferenti 8–15 ramos umbelliformes parvos; floribus 40–60 mm longis viridi-luteis; ovario 20–30 mm longo fusiformi apice constricto sulcato; tubo 5–10 mm longo 10–13 mm lato infundibuliformi; segmentis subequalibus luteis 15–20 mm longis 3.5–5 mm latis, linearibus involutis apiculati-cucullatis; filamentis 35–40 mm longis ad apice tubi inaequaliter insertis; antheris 15–20 mm longis, luteis; capsulis 3.5–4.2 cm longis 1.5–2 cm latis, oblongis stipitatis bruneis; seminibus 5–6 mm longis 4–4.5 mm latis lunatis vel lacrimiformibus margine alatis undulatis.

Relationships and Comparisons.—*Agave deserti simplex* is distinguished from *A. deserti deserti* by its predominantly solitary habit, deeper flower tube, tepals being only ca. 2–3 times longer than the tube (vs. 4–5 times longer), filaments

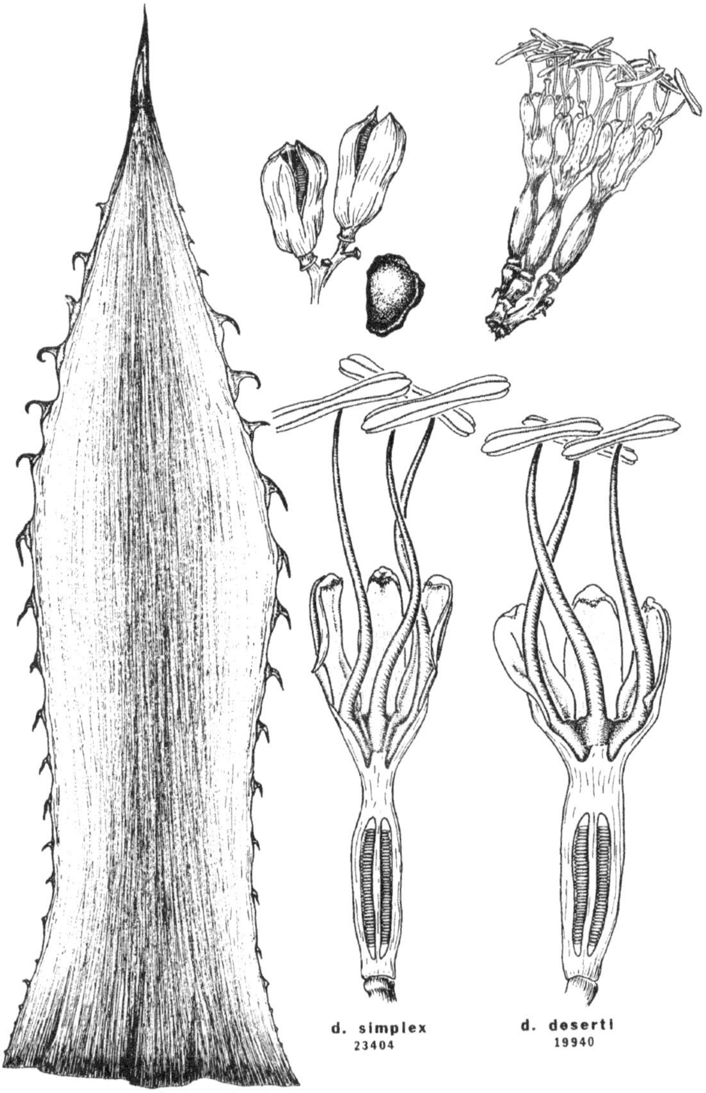

FIGURE 11. *Agave deserti* ssp. *simplex* drawn from type, *Gentry 23404*: leaf, flower cluster, capsule ×½; flower section ×1½; seed ×2. *A. deserti* ssp. *deserti*, flower section from *Gentry 19940*.

FIGURE 12. *Agave deserti simplex* in the Silver Bell Mountains, Pima County, Arizona. The plants are frequently depauperate, resemble *A. mckelveyana* with small flowers, 33–46 mm long, but the tube is characteristically deep, 5–7 mm.

being inserted in the tube (vs. inserted on the base of the tepals), and the disjunct distributions of the two taxa (Fig. 7). It is not possible to separate the two taxa on leaf specimens alone, even when accompanied by photographs. However, if labels state the origin of the specimen and whether the population from which it was selected consists of cespitose plants or generally of singles, identification of leaf specimens becomes relatively certain. Large plants frequently offset, especially at maturity, but the offsets may wither and die.

Another close relative of *A. deserti simplex* is *A. mckelveyana*, which occupies higher montane elevations northeast of the *simplex* area. As in *A. deserti simplex*, the filaments of *A. mckelveyana* are inserted in the orifice of the tube, but the latter is a smaller plant with small narrow leaves, slender wandlike inflorescence, and small flowers. It is possible that these two taxa will be found in contact, and if so, it will be interesting to see if there is introgression.

In Arizona the scattered populations of *simplex* rosettes sucker much more sparingly than the California colonies, and the large cespitose California clones are absent. In the Lechuguilla Desert of southwestern Arizona, one of our most arid regions, the subspecies is represented by small drought-depauperized, scattered individuals, with leaves commonly only 20–25 cm in length. Other than for their small size, the characters of the plants are quite like those of

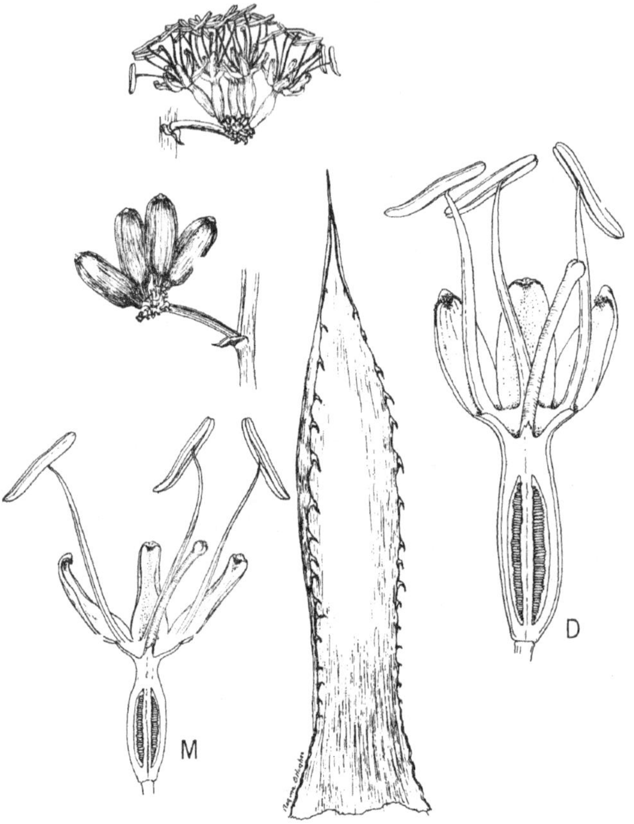

FIGURE 13. (*M*), *Agave mckelveyana* from type: leaf, flower cluster, and capsules ×¼; flower section slightly enlarged. (*D*), *A. deserti* flower section slightly enlarged.

the species—rather broad, gray, cross-banded, thick rigid leaves and slender small panicles. However, also found in the region is a long-leaved robust form in the Mohawk Mountains (*Wiggins 8643*), which I have not seen in habitat.

The native uses of *Agave deserti simplex* are essentially what has been reported above for the species.

Agave mckelveyana

(Figures 7, 13–15; Table 2)

Agave mckelveyana GENTRY, Cactus Succulent J. (U.S.) 42:225. 1970.

Small, single or suckering, rather few-leaved rosettes 20–40 cm tall; leaves 20–35 × 3–5 cm, linear or lanceolate and broadest in the middle, light glaucous green or yellowish green, firmly spreading, the margin nearly straight or undulate with low teats; spine 1.5–4.0 cm long, shortly decurrent, subulate, rounded except for shallow groove above at base, castaneous to gray; teeth small to medium, the larger at mid-blade 4–8 mm long, mostly 1–3 cm apart and downflexed, grayish with reddish tips, rather friable; panicle small, narrow 2–3 m tall with 10–19 laterals in

FIGURE 14. *Agave mckelveyana* near Sitgreave Pass in the Black Mountains, June 1966, with highland dispersed desert shrub at ca. 3,500 ft (1,070 m) altitude, and Linnea Gentry.

upper half of shaft with small compact umbels; pedicels short-bracteolate; flowers small, 30–40 mm long, the perianth and stamens openly spreading, yellow; ovary light green, 16–22 mm long including constricted neck faintly grooved below tepal sinuses, fusiform or cylindric; tube shallow, open, 3.0–4.5 mm deep, 8–9 mm broad; tepals 12–13 × 3–4 mm spreading, thin, with open sinuses, linear conduplicate, the outer longer and narrower than the round-keeled inner, both abruptly hooded at tips; filaments 25–30 mm long, inserted in orifice of tube, flattened towards base; anthers 12–16 mm long, yellow; capsules 30–45 × 10–14 mm, oblong, narrowly stipitate, obtuse to apiculate, thin-walled, striate; seeds 5.0–6.5 × 4.0–4.5 mm, obovate from apical hilum or half-moon, the faces rugose, completely margined with a low wing.

Type.—Gentry 21979, Sitgreave Pass in Black Mountains, ca. 4 miles [ca. 6 km] NE of Oatman, Arizona, 26 June 1966,

US. It occurs here as a small scattered population on rocky volcanic slopes between 3,000 and 4,000 ft (900 and 1,200 m) elevations.

Agave mckelveyana occupies a central part of western Arizona (Fig. 7). Its habitat is with the chaparral and juniper associations on rocky slopes between 3,000 and 6,000 ft (ca. 900 and 1,800 m) elevations. I have not seen it growing with other *Agave* species, but McKelvey collected it in contiguous numbers with *A. utahensis* in the Aquarius Mountains and with *A. parryi* var. *couesii* in the Juniper Mountains. It appears to occupy an ecologic niche of its own, quite distinct from that of its near relative, *A. deserti simplex*, which is confined generally to the Sonoran Desert at lower elevations. The habitat of *A. mckelveyana* is better watered and is cooler than that of *A. deserti*.

Marcus Jones collected a cutting of inflorescence in 1930 near Oatman (*Jones 25167*) and mixed it with his specimens which he described as *A. utahensis* var. *discreta*. Trelease annotated some of McKelvey's earlier collections as *A. aquariensis*, but the name was never published. The slender panicles, small flowers with shallow tubes and spreading tepals place it in the section Deserticolae of Trelease. Its nearest relatives appear to be *A. deserti* and *A. subsimplex* Trel. The small leaves of *A. mckelveyana* with variably flexed teeth resemble those of *A. subsimplex*, from which it can hardly be separated except by the flower differences. The flowers of *A. mckelveyana* differ from these two, and other relatives, in smaller size, in the narrow spreading, thin, linear, conduplicate tepals with sharply inflexed cucullate tips, and small ovary with constricted neck.

Some observers of *A. mckelveyana* have regarded it as a depauperate form of *A. deserti*. The plants in some of the drought-inhibited populations of *A. deserti simplex*, as in the Lechuguilla Desert, are about the size of *A. mckelveyana*, but they still have the essential, if subtle, characters of *A. deserti*. The latter are generally more robust than *A. mckelveyana*. The pattern of variation in *A. mckelveyana* is phenotypically homogenous and does not intermesh with the variability of the polymorphic *A. deserti*. A leaf series of *A. mckelveyana* is shown in Fig. 15; flowers are shown in Fig. 13. I find no intergrading between the two groups and believe them to be biologically distinct species. They probably separated from a common ancestor long ago.

There is nothing particularly remarkable about this small *Agave mckelveyana*, and its habit of growing among shrubbery keeps it obscure. The flowering season appears to be during May, June, and July. However, flowering time will vary from year to year and according to elevation. The life span is not known. Mature flowering and fruiting rosettes observed had only 30–40 leaves, but as the rate of leaf growth is unknown, the relatively few leaves do not necessarily indicate a short-lived rosette. Suckering offsets were observed on other plants on the Black Mountains and again on Hualapai Mountain.

In the Hualapai Mountains, Abert's tree squirrel, *Sciurus abertii*, has been observed gathering green capsules (word of Rodney Engard). In one instance the squirrel was observed cutting and carrying the fruits down. Other fruit-denuded panicles in the same locality were attributed to the work of this animal.

Agave subsimplex

(Figures 7, 16, 17; Table 2)

Agave subsimplex TREL. Missouri Bot. Gard. Rep. 22:60. 1912.

Small, single or cespitose, glaucous, colorful, low-spreading rosettes, 20–35 cm tall, 50–70 cm broad; leaves variable, 12–35 × 3–5 cm, lanceolate to ovate, long-acuminate to short-acuminate, but little narrowed towards the base, thick, rigid, rounded on back, hollowed in inner face, gray glaucous or light yellow-green, or sometimes purple-tinged, the margin nearly straight or strongly teated; teeth variable, friable, the larger 3–15 mm long, straight or variously flexed, rarely bicuspid, brown or more often yellowish gray; spine subulate, 2–4 cm long, not or but little decurrent, frequently sinuous, shallowly grooved above, glaucous gray; panicle slender, narrow, 2–3.5 m tall, with 5–8 short laterals; flowers in small umbels, yellow to pink 40–45 mm long; ovary about 25 mm long with an unconstricted long (5 mm) neck; tube shallow, spreading, 3–4 mm deep, 10 mm wide; tepals 12–15 × 6–7 mm, equal, ascending, elliptic, plane, widest at the middle, apiculate and scarcely hooded; filaments 25–28 mm long, round in cross section, inserted below base of tepals 3 mm above bottom of tube; anthers 13–

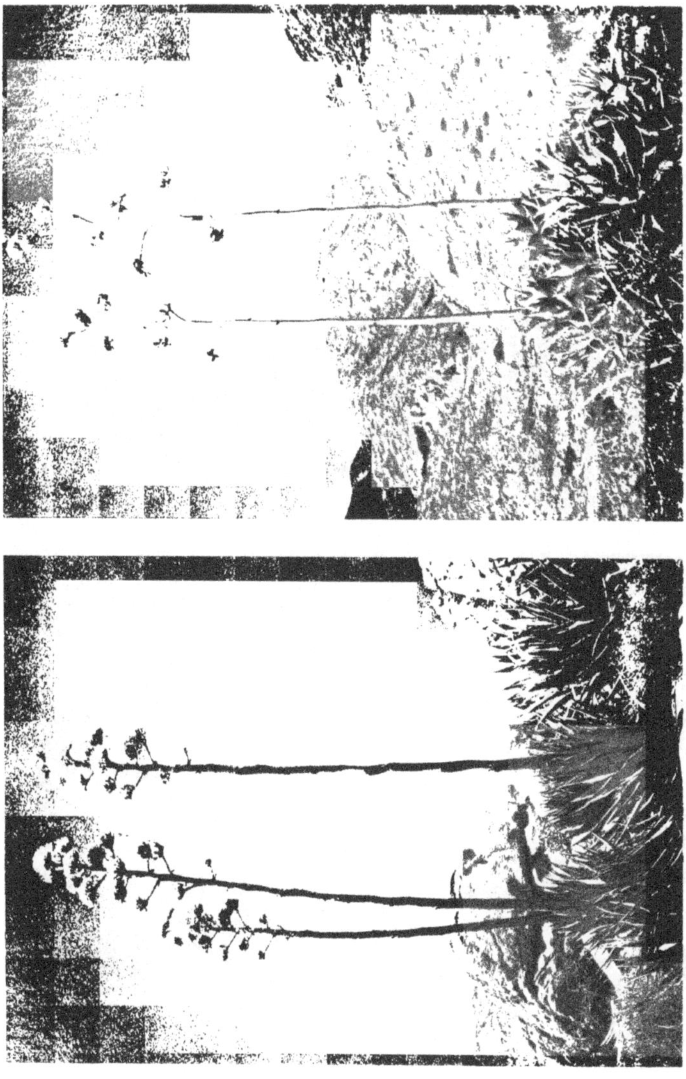

PLATE 1. (*Left*) *Agave deserti*. La Rumerosa. Photograph by George E. Lindsay, June 1960. (*Right*) *Agave margaritae*. Bahía Santa María. Photograph by Reid V. Moran, 31 March 1952.

PLATE 2. (*Top*) *Agave gigantensis* on the volcanic top of Sierra de las Palmas. Photograph by the author, April 1952. (*Bottom*) *Agave cerulata dentiens*, Isla San Esteban. Photograph by Reid V. Moran, 17 April 1975.

FIGURE 15. *Agave mckelveyana* in the Black Mountains. (*Upper*) Mature leaves from 6 plants, the leaf at right depauperate from a poor site, the leaf at left from the rosette shown below.

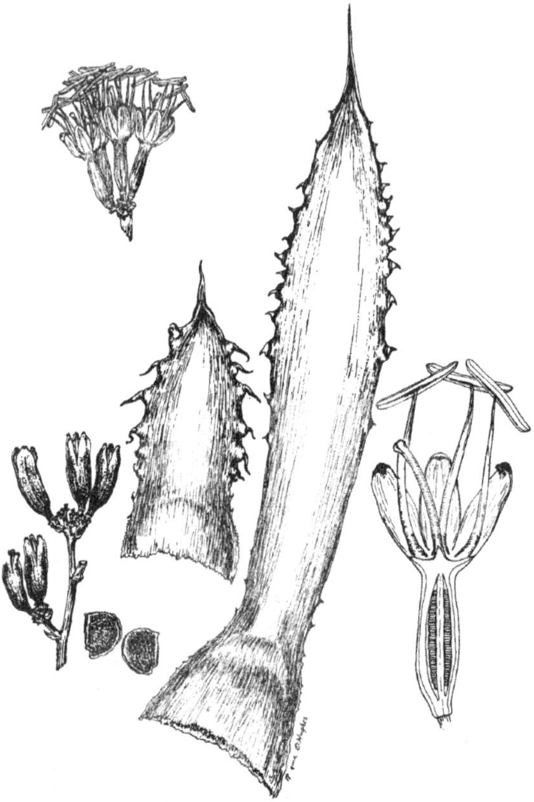

FIGURE 16. *Agave subsimplex*: larger leaf drawn from *Gentry 10217*, the smaller from *Gentry 10221*; flowers from *Felger & Bezy 14171*, section ×1, cluster ×2/5; capsules ×1½, and seeds ×1¾ from *Gentry 10221*.

15 mm long, centrically attached; capsules 40 × 15 mm to 35 × 10 mm, variable, oblong, sometimes narrowly so, light glaucous (*Gentry 10221*), bluntly apiculate, narrowly or broadly stipitate, the valves thick and striate-nerved; seeds (*10221*) mostly 4.5 × 3 mm, sooty black, roughly lunate, hilum notch narrow and the opposite corner frequently apiculate, edges rimmed with a sharp, winglike flange.

Type.—SONORA: Seal Island, just off Tiburon Island, 13 Apr. 1911, *Rose 16811*, US. Known only from coastal Sonora, where it is thinly scattered in small colonies on the outwash slopes of the granitic and volcanic mountains and the adjacent islands.

Reid Moran in 1966 found *Agave subsimplex* common on Turner's Island, or Isla Dátil, as it is also called, just south of Tiburon Island in the Gulf. Moran's (1967) account provides a short botanic history of the plant, a diagnostic description, and his field observations.

This small xerophytic agave is closely related to *A. deserti* and *A. cerulata* Trel., as expressed by their common characters of small variable leaves, nearly tubeless flowers, and narrow

PLATE 3. (*Left*) *Agave cerulata dentiens* in natural cluster and habitat on Isla San Esteban. Photograph by Reid V. Moran, 6 May 1952. (*Right*) *Agave promontorii*. La Aguja. Photograph by Reid V. Moran, 18 May 1959.

PLATE 4. (*Top*) *Agave aurea* 19 km southeast of La Paz. Photograph by Reid V. Moran, 25 January 1959. (*Bottom*) *Agave aurea* "southwest end of Mesa de San Geronimo, northerly from Rancho Viejo (on road from Loreto to San Jávier). Alt. ca. 1,100 m. Lat. 20°58′N, long. 111°32.5–34′W." Photograph and quotes by Annetta Carter, 8 May 1966.

FIGURE 17. *Agave subsimplex* along the arid Sonoran coast, Feb. 1951.

small panicles, which characterize the group Deserticolae. I think the characters outlined in the key to species should be sufficient to distinguish *A. subsimplex* from *A. deserti*, but depauperate plants of the latter in the Lechuguilla Desert of Arizona resemble *A. subsimplex*. The two species are not known to cohabit, but it is possible that they may be found together on some unexplored mountains in northwestern Sonora, such as the coastal Sierra del Alamo. *Agave subsimplex* may be more closely related to the polymorphic *A. cerulata* of Baja California, as indicated by the small, variable leaves and the narrow, oblong, waxy capsules. However, the Sonoran plants have more spreading rosettes, gray not yellow, with thicker, wider, less acuminate leaves, and the filaments are inserted below the base of the tepals. In *A. subsimplex* the nectarious innerliner of the tube, from the rim of which the filaments ascend, does not completely fill the tube, as it does in *A. cerulata*. The flowers of *A. subsimplex* are smaller than those of *A. cerulata*, 40–45 mm vs. 45–60 mm. The pink to red color frequently occurring in the pistil, filaments, and corolla (see notes of collectors in Exsiccatae) is distinctive of *A. subsimplex*. Such color is not known in the flowers of *A. deserti* and *A. cerulata*.

The Seri Indians, a maritime hunting-and-gathering people of the Sonoran coast, called this agave "ahmmo," as nearly as I could render the sound, and stated that it was gathered for cooking and eating. Whiting noted the Seri name as "den; kl." Small amounts of sapogenin (0.07–0.14 percent) were found in the leaves. Felger and Moser (1970) render the name phonetically as "?aamXW." They reported that the pit-baked stems are eaten in several ways: cubed and cooked with sea turtle, or made into flat cakes or patties, which may be stored for later use or taken on long trips. These cakes could also be soaked in water and consumed as a sweet drink. The various agave viands all had Seri names and indicate an habitual, important resource for survival by a people living on the outer edge of human environment.

Agave cerulata

Agave cerulata Trel., ssp. **cerulata**

(Figures 18–21; Tables 2, 4)

Agave cerulata TREL. Missouri Bot. Gard. Rep. 22:55. 1912.

Small yellow to pale green, rarely light glaucous gray, abundantly surculose, few-leaved rosettes 25–50 cm tall; leaves mostly 25–50 × 4–7 cm, long acuminate, narrowly lanceolate to triangular-lanceolate, yellow to light green, sometimes cross-zoned, the margins nearly straight to mildly undulate with teated teeth; teeth small, 1–4 mm long, irregularly spaced, sometimes lacking through much of the blade, grayish brown, bordered with a brown ring at base, weakly attached; spine 3–6 cm long, acicular, light gray to dark gray, decurrent only to uppermost teeth or less; panicle slender narrow, 2–3.5 m tall, with mostly 6–12 small lateral umbels, the whole white waxy glaucous in bud stage; bracts small scarious triangular; flowers waxy white in bud, opening pale yellow, mostly 45–60 mm long; ovary 22–32 mm long, fusiform narrowly tapered toward base; tube 3 × 11 mn to 5 × 14 mm, broadly funnelform or discoi with thick nectary and prominent bulges oppo site filament insertions; tepals 16–22 mm long ascending to spreading, equal, elliptic, the inne wide with low broad keel and broadly over lapped at base by outer; filaments inserted a base of tepals on rim of nectary, variable i length, usually incurved at anthesis, 30–40 mn long; anthers 15–20 mm long, centric to excen tric; capsules 3–5 × 1.2–1.3 cm, narrowly ob long, waxy light gray, narrowly stipitate, bluntl apiculate; seed 5 × 3 mm, lunate, sooty black the hilar notch open or obscure, the margina wings pronounced on both sides around the cur vature.

Type.—*Nelson & Goldman 7180*, Calmallí, Baja Californi 29 Sept. 1905, US.

Characteristics and Habitat.—Most plants of *A cerulata* are characterized by slender, yellow long-acuminate, lanceolate leaves with brow eyelets ringing the weakly attached, moderat to small teeth, small narrow panicles whitene from peduncle to capsules with a waxy bloom and light-yellow spreading tepals. *A. cerulat* has been confused with its near relative *A. de serti*, but the latter is more robust, the leave light gray-green rather than yellow-green, 4–times longer than broad (vs. 5–12 times longe than broad). *A. deserti* generally lacks the dis tinctive brown eyelets ringing the teeth and th capsules are generally broader and, althoug glaucous, lack the waxy white bloom. The ge graphic distributions of the two species are dis tinct (cf. Figs. 7 and 21), although they may mee on the peninsula south of Sierra San Pedro Má tir in an area I have not explored.

Agave cerulata is primarily a product of th upland maritime environment where freque fogs and ocean breezes temper the desert cl mate, while *A. deserti* is a product of the h arid continental desert with high insolation an more extreme circadian temperature changes Figures 9 and 20 show the two species in thei characteristic respective habitats. There are sig nificant chemical differences between these tw species, as may be inferred from Tables 3 an 4.

Chemistry.—The data in Table 4 on the sap genin contents of *Agave cerulata* are the mo detailed for any *Agave* species and is assemble

TABLE 4. SAPOGENIN CONTENT IN *Agave cerulata*. Collated from Wall (1954) and Wall et al. (1954a, 1954b, 1955, 1957). Sapogenins are given in percentages on dry-weight basis: hec. = hecogenin; git. = gitogenin; man. = manogenin; tig. = tigogenin; fl. = flower; dead = leaf of dead plant.

				Sapogenins				
Coll. no.	Baja California source locality	Month coll.	Plant part	% total	% hec.	% git.	% man.	% tig.
	Agave cerulata cerulata							
10346	Calmallí	May	leaf (white)	1.2	25		68	7
10346	Calmallí	May	leaf (yellow)	1.1	65		35	
10346	Calmallí	Dec.	leaf (white)	1.1	X	X	X	X
10346	Calmallí	Dec.	leaf (yellow)	0.5	X	X	X	
10346	Calmallí	Dec.	leaf (dead)	1.2	X		X	
11322	Laguna Chapala	Apr.	leaf	1.1	20		45	X
11322	Laguna Chapala	Dec.	leaf	1.0	X	X	X	X
11322	Laguna Chapala	Dec.	leaf (dead)	0.5	X		X	
11322	Laguna Chapala	Dec	leaf	0.4	50		50	
11322	Laguna Chapala	May	leaf	1.2	35		65	
11324	Laguna Chapala	Apr.	leaf	0.1	X		X	
11188	Laguna Chapala	Sep.	leaf	0.8	69	31		
11740	Sierra San Luis	Apr.	leaf	0.6	80			20
11741	Sierra San Luis	Apr.	leaf	0.72	60		40	
11744	Sierra San Luis	Apr.	leaf	0.9	7			93
11746	Sierra San Luis	Apr.	leaf	0.6	20		70	10
11919	Calmallí	May	leaf	0.6	40		60	
11926	Calmallí	May	leaf	0.6	45		55	
11924	Calmallí	May	leaf	0.8	55		25	20
11962	Laguna Chapala	May	leaf	0.1	100			
11962	Laguna Chapala	Dec.	leaf	3.4	X	X	X	X
11953	E of Punta Prieta	May	leaf	0.4	30		70	
	Agave cerulata nelsonii							
10370	San Fernando	Sept.	leaf	1.2			100	
10370	San Fernando	Sept.	stem	0.2	70		15	
10370	San Fernando	Apr.	leaf	0.7	5			95
10370	San Fernando	Apr.	leaf	0.25				100
10370	San Fernando	ec.	leaf	1.3			100	
10370	San Fernando	Dec.	stem	0.6			100	
10370	San Fernando	Dec.	leaf (dead)	1.0	X	X	X	X
10370	San Fernando	May	leaf	1.1	X	X	X	X
11160	San Fernando	Sept.	leaf	0.77	75		25	
11160	San Fernando	Apr.	leaf	0.94	70		30	
11162	San Fernando	Sept.	leaf	1.0	90		10	
11162	San Fernando	Apr.	leaf	0.6	50		30	20
11164	San Fernando	Apr.	leaf	0.5	75		25	
11170	San Fernando	Dec.	leaf	2.2	X		X	X
11665	San Fernando	Apr.	leaf	0.3			50	50
11665	San Fernando	Dec.	leaf	2.5	X		X	X
11665	San Fernando	Dec.	leaf	1.3	X		X	X
11666	San Fernando	Dec.	leaf	0.5	55			
11669	San Fernando	Dec.	fl.	0.25		60		
	Agave cerulata subcerulata							
10330	San Ignacio	Apr.	leaf	0.35	64	8	11	17
10330	San Ignacio	May	leaf	0.3	45	30		25
10330	San Ignacio	Dec.	leaf	1.5	X	X	X	X
11892	Isla San Marcos	May	leaf	0.7	X		X	
12393	E of San Ignacio	Dec.	leaf (dead)	2.0	X		X	

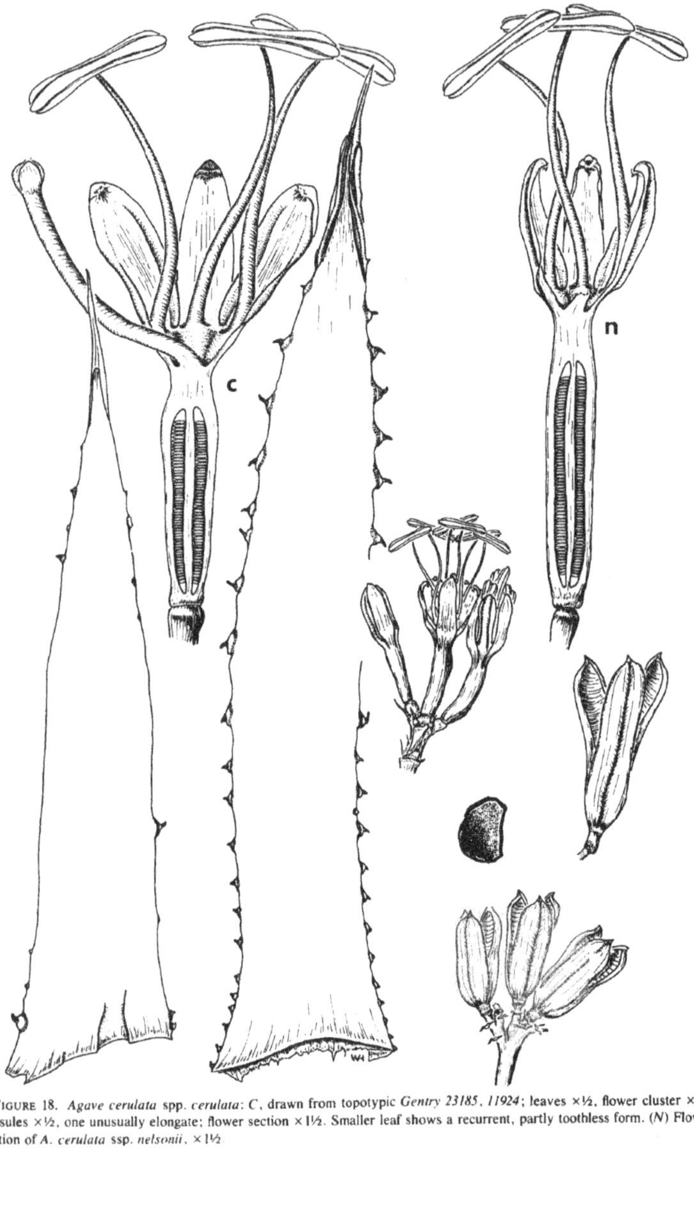

FIGURE 18. *Agave cerulata* spp. *cerulata*: *C*, drawn from topotypic *Gentry 23185, 11924*; leaves ×½, flower cluster ×3/5, capsules ×½, one unusually elongate; flower section ×1½. Smaller leaf shows a recurrent, partly toothless form. (*N*) Flower section of *A. cerulata* ssp. *nelsonii*, ×1½.

PLATE 5. (*Top*) *Agave shawii*. Inflorescence of plant photographed about 16 km southeast of El Rosario, Baja California. Photograph by E. S. Ross, 17 November 1977. (*Bottom*) *Agave shawii* with *Simmondsia chinensis* along the Pacific shore north of Ensenada, spring 1952. Many such stands have been cleared away since that date. Photograph by the author.

PLATE 6. (*Left*) *Agave shawii* spp. *goldmaniana* near Punta Prieta. Photograph by George E. Lindsay, May 1975. (*Right*) *Agave sebastiana* on West San Benito Island. Photograph by George E. Lindsay, February 1978.

FIGURE 19. (*Upper*) *Agave cerulata cerulata* near Laguna Seca Chapala. The dense clusters of rosettes with long acuminate leaves are characteristic. (*Lower*) *A. cerulata cerulata* east of Punta Prieta along road to Bahía de Los Ángeles, May 1952; a robust, almost white-glaucous form with cross-zoned leaves.

FIGURE 20. (*Upper*) Fairy ring of *Agave cerulata cerulata* on calcareous pebble pavement in northwest corner of San Andres Valley, spring, 1973. Bruce Gentry photograph. (*Lower*) *A. cerulata cerulata* on rolling plain near Laguna Seca Chapala, May 1952.

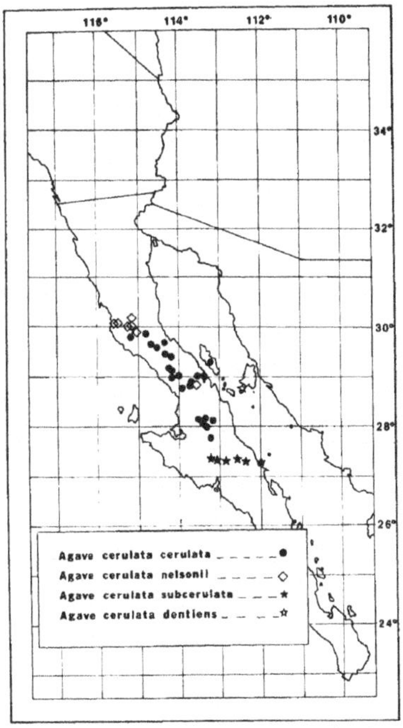

FIGURE 21. Distribution of *Agave cerulata* and its subspecies based on herbarium specimens, see Appendix—Exsiccatae.

from U.S.D.A. sources (Wall 1954; Wall et al. 1954a, 1954b, 1955, 1957). Of the four genins present, hecogenin and manogenin are the more prevalent. Hecogenin occurs in 89%, manogenin in 82%, tigogenin in 24% of the 46 plant samples. These four compounds shift about quite at random, as there is little apparent correlation with time of year, locality, plant part, or stage of plant maturity. The three cases in the table of "leaf, dead" represent dry leaves from rosettes that had flowered and died. However, it is apparent that total sapogenin content is higher in December than during other months. The average content of the 13 leaf samples of December is 1.5%, or about double the average content of 0.66% of the 28 other seasonal leaf samples.

The average sapogenin content of *Agave cerulata cerulata* is 0.87%, of *A. cerulata nelsonii* is 0.96%, and of *A. cerulata subcerulata* is 0.98%. This is nearly 1% and contrasts strongly with 0.25% for *A. deserti*. Although samples of the latter are few, they show that there are chemical and physiological differences between *A. deserti* and *A. cerulata* as well as morphological and geographic ones.

Populations.—The populations of *Agave cerulata* are vast, extending through northern and central Baja California from just above latitude 30°N to ca. 27°N (Fig. 21). The clustering clones must number over a million and the individual rosettes many millions more—a wild species population probably exceeded only by *Agave lechuguilla* of the northern desert region of Mexico. *A. cerulata* is polymorphic, showing several forms at many localities. An example of this was recorded on 17 May 1952 near Calmallí, the type locality. The forms noted here were itemized as follows, with Gentry collection numbers in parentheses:

A. Rosettes large, leaves 45–60 cm long; flg. stalk to 6–7 m (*11919*).
B. Size medium, glaucous-yellow leaves (*11921*).
C. Size small, leaves thick (*11922*).
D. White glaucous with slender narrow leaves (*11923*).
E. Thick-headed glaucous rosettes with part of the margins toothless (*11924*).
F. Short, thick-leaved, yellow-green with numerous proximal teeth (*11926*).

Corresponding forms as well as other striking variants were noted in other localities. The extensive collections and detailed study necessary to organize a rational taxonomy of these genetic forms has not been attempted. However, the three geographic subspecies set forth below will clarify some relationships.

Agave nelsonii Trel. has long been considered conspecific with *A. deserti* (Johnston 1924; Shreve and Wiggins 1964) and indeed is very similar in habit, in the gray glaucous leaves, and in other characters. However, the ceriferous inflorescence of *A. nelsonii*, the brown eyelets ringing the teeth, its chemistry and its geographic distribution relate it with the *A. cerulata* complex. I am accordingly treating it as a subspecies of *A. cerulata*. This and two other variants are separable according to the following key and descriptions.

Key to Subspecies of *Agave cerulata*

1a. Leaves yellowish or light glaucous, long-acuminate, mostly 6–12 times as long as broad, the margins nearly straight to mildly undulate; teeth small, fragile; spines acicular 3–6 cm long 2

1b. Leaves mostly light-gray to bluish glaucous over green, short-acuminate, mostly 3–6 times as long as broad, the margins undulate to mammillate; teeth larger, firmer; spines subulate, 2–4 cm long 3

2a. Leaves generally yellowish, 25–50 cm long; panicles narrow, the lateral branches 15–20 cm long on upper ⅓ to ¼ of shaft ...... *c. cerulata*

2b. Leaves light glaucous gray, 40–55 cm long; panicles broad, the lateral branches 30–40 cm long on upper ½ of shaft. Isla San Esteban ...... *c. dentiens*

3a. Leaves larger, 25–40 cm long, the margins undulate with small teats or nearly straight. Sierra San Miguel to San Agustín ...... *c. nelsonii*

3b. Leaves smaller, 15–28 cm long, the margins conspicuously crenate with prominent teats. Vicinity of San Ignacio to San Marcos Island ...... *c. subcerulata*

Edibility of A. cerulata.—Considering the relatively high sapogenin contents, *Agave cerulata* is probably not among the best for eating. However, it was eaten by the Cochimi Indians, a fairly large tribe inhabiting the central part of the peninsula. They ate the thick stem or *cabeza*, which contains less sapogenins than the leaves. Wenceslaus Linck, a Jesuit missionary, founded the Misión de San Francisco Borja during the 5th decade of the 18th century. In 1766 he set out with thirteen soldiers and a band of missionized Indians to explore northern Baja California to the mouth of the Colorado River (Burrus 1966). He traveled over 480 km through the heartland of *Agave cerulata*. Of the second day northwest of San Borja at Vimbet he wrote, "A native chieftain and his entire settlement of Nuestra Sonora de Guadalupe had prepared for us thousands of *mescales*, a food which is their daily bread and their favorite sustenance" (ibid.:45).

He mentioned *mescales* several times as the most common food the Indians gave to them along his trail, in exchange for his friendship and blessings. When crossing the southern end of the Sierra San Pedro Mártir on March 13, he noted, "Our great pagan friend regaled us here with ten loads of ground *mescal*. Many unconverted natives gathered here, accompanied by a pagan savage whom I cured." His account is the first we have on these wild people of northern Baja California and leaves no doubt of the importance of agave as food in this desert land. Further confirmation of this is given for more southern Indians in the wonderful account written by Padre Barco of the Misión de San Javier in the Sierra de la Giganta (see Introduction).

Today, except in the Sierra San Pedro Mártir region, there are no Indians left to eat the agaves. However, cattle, including horses, burros, cows, and goats, crop the young flowering shoots extensively. Horses will bend over the flowering stalks and eat the flowers. Near Tres Virgines Mountain, I observed where some large animal had learned to break out the terminal bud of *A. cerulata subcerulata* and eat the soft central tissue of stem, bud, and leaf bases. The young flowering stem contains little or no sapogenin. The heads or stems of *A. cerulata dentiens* were eaten by the Seri Indians, who plied across the Gulf to reach them (see below).

Agave cerulata ssp. **dentiens** (Trel.) Gentry, stat. nov.

(Figure 21; Tables 2, 4)

Agave dentiens Trel. Missouri Bot. Gard. Rep. 22:51. 1912.

Medium-sized, acaulescent, surculose, open, few-leaved, green to light glaucous gray rosettes, 50–70 cm tall, 8–15 dm wide, forming dense clumps; leaves 40–70 cm long, 4–7 cm wide in the middle, wider below, triangular-long-lanceolate, thickly crescentic at base, concave above towards apex, rigid, light gray glaucous, sometimes bluish, cross-zoned, the margins straight, usually set with small, weak, friable teeth 1–2 mm long, or margin nearly toothless; spine acicular, 3–5 cm long, brown to gray, narrowly, shortly channeled above, shortly decurrent; panicle 3–4 m tall, open, with 8–18 wide-spreading lateral branches in upper half of shaft; buds ceriferous, pale yellow, in small congested umbels; flowers pale yellow, slender, 49–53 mm long; ovary fusiform, 32–35 mm long with slender neck, constricted at base; tube shallow, open, 3 × 10 mm, swollen with nectaries below filament insertions, grooved; tepals ca. equal,

15–16 mm long, 4–5 mm wide, ascending-spreading, occasionally recurving, conduplicate and involute with anthesis, rounded and shortly hooded at apex; filaments slender, 28–30 mm long, bulbous at base with nectaries; anthers 15–16 mm long, yellow, centric; capsules glaucous ceriferous, 40–50 × 15–20 cm, oblong, short-stipitate, short-beaked. (Flower description from *Felger s.n.* San Esteban Island, 20 June 1977).

Type.—Rose 16819, San Esteban Island, Baja California 12 Apr. 1911, MO, US.

There is little doubt that *Agave dentiens* is closely related to *A. cerulata*. The old frayed capsules on the type specimen show a waxy bloom like that of *A. cerulata*, while the long, narrow, acuminate leaf-form and small, weak, brown-eyeleted teeth are additional evidence for *cerulata* affinity. The random occurrence of toothless margins is also a *cerulata* character. A long-leaved, light-glaucous variant (*Gentry & Fox 11953*) of *cerulata* (Fig. 19) found on the peninsula between Punta Prieta and Bahía de Los Ángeles looks like the island *dentiens*. The satellitic-island position of *dentiens* in relation to the peninsular *cerulata* populations is very similar to that of *A. sobria roseana*, here also treated as a subspecies. Moran's color photographs of collection *4079* on San Esteban Island show a generally uniform population with rather consistent characters of a wide-branching panicle and tall, long-leaved, light gray-glaucous, narrow rosettes. This is in obvious distinction from the narrow panicles and usually small, yellowish rosettes of peninsular *cerulata*. But this difference seems to be at the subspecific level. It is possible that the agaves on Angel de la Guarda Island should be aligned with *A. cerulata dentiens*, but until flowering specimens of the respective populations are assembled, the disposition would be uncertain. Altogether, ssp. *dentiens* looks like an isolated genome undergoing evolution and, perhaps, on its way to complete speciation.

The Seri Indians called *Agave cerulata dentiens* "ʔemme" (Felger and Moser 1970). The Seri visited San Esteban Island to collect the plant and found the more fibrous plants the most savory, distinguishing also a white-and-green variety. Large baking pits were observed on the island, and the Seri reported that as many as 100 heads were cooked at a time. They also relieved thirst with juice of the macerated leaves after they were cooked in a fire. The juice was also drunk after fermenting for several days with warm water added. Felger reports (personal communication) that the green-leaved form generally occupies the higher elevations of the island, where fogs or clouds tend to hang and reduce insolation.

Agave cerulata Trel. ssp. **nelsonii** (Trel.) Gentry, stat. nov.

(Figures 21–23; Tables 2, 4)

Agave nelsonii Trel. Missouri Bot. Gard. Rep. 22:61. 1912.

Short-stemmed, rather small, cespitose, compact, glaucous green rosettes, 50–75 cm in diameter at maturity; leaves mostly 20–35 × 6–8 cm, rarely larger, lanceolate to triangular-lanceolate, short-acuminate, usually a little narrowed above a broad base, thick rigid, light green with a gray to bluish glaucous bloom; teeth in the mid-blade 3–9 mm long, frequently on small teats, brown-ringed at base, grayish brown, closely regular-spaced 1–2 cm apart, relatively firmly attached; spine 2–4 cm long, strongly subulate, flat or openly grooved above, grayish brown, decurrent to first or second pair of upper teeth; inflorescence slender, 2.5–4 m tall, with 15–20 ascending to arching laterals in upper half of shaft, the yellow flower umbels compact, globose; flowers slender, light yellow, 45–55 mm long; ovary 25–35 mm long, slightly angled, the neck long, grooved, slightly constricted; tube small, 3–5 mm deep, openly funnelform; tepals subequal, linear, erect, incurved at tip, hooded, the inner with prominent keel; filaments slender, 30–40 mm long, inserted at base of tepals on nectary disk; anthers centric, 15–18 mm long; capsules 4–4.5 × 1.5 cm, oblong or smaller and pyriform, stipitate, with acute beak; seeds 4 × 5 mm with hilar notch at apex and thin marginal wing.

Type.—Nelson & Goldman 7111, San Fernando (Sierra San Miguel), Baja California, 4 Sept. 1905, US. Goldman's photo (Trelease 1912:pl.65) shows very well the habit of this plant and its habitat on the igneous highlands with open cover of cirio, ocotillo, cactus and low shrubs.

Agave cerulata Trel. ssp. **subcerulata** Gentry, ssp. nov.

(Figures 21, 23 (*lower*), 24, 25; Tables 2, 4)

Small, acaulescent, few-leaved, open, glaucous green cespitose rosettes 15–30 cm tall, 30–

FIGURE 22. *Agave cerulata nelsonii* on igneous rocky slope of Sierra San Miguel near San Fernando, 22 June 1973. Photograph by John McClure.

50 cm in diameter; leaves 15–30 × 2.5–7 cm, lanceolate to triangular lanceolate, thickly fleshy, arching upward, guttered at maturity, white glaucous to glaucous green, margins crenate with prominent tests under well developed teeth; teeth at mid-blade 3–8 mm long, variously flexed, weakly attached, grayish brown, 1–3 cm apart; spine 3–4 cm long, subulate, usually sinuous, grayish brown; panicles 2–3 m tall, slender, waxy glaucous, with mostly 8–10 small lateral umbels of light yellow flowers; flowers 44–55 mm long; ovary 23–30 mm long with grooved neck 4–7 mm long; tube 3–6 mm long, shallow, openly spreading; tepals equal, 17–22 × 6–7

FIGURE [illegible]. (*Upper*) *Agave cerulata nelsonii*, detail of rosette showing the typical short-acuminate leaves. (*Lower*) *A. cerulata subcerulata* near San Ignacio; the few-leaved rosettes with short thick leaves are characteristic.

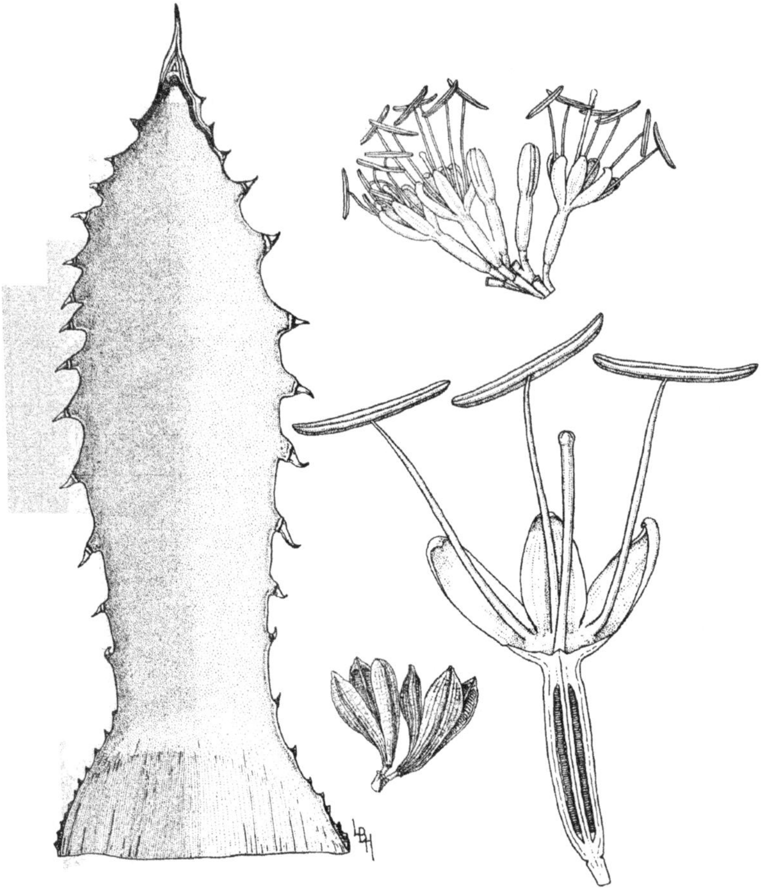

FIGURE 24. *Agave cerulata* ssp. *subcerulata*, leaf drawn from type *Gentry 10330*; flowers from topotypic *Gentry 23170*. Leaf, flower cluster, capsule ×½; flower section ×1½.

mm, smooth, thick, oblong, apex broadly rounded, slightly hooded, the inner with broad low keel; filaments 33–40 mm long, inserted on rim of nectary disk at base of tepals; anthers 18–21 mm long; capsules waxy glaucous, mostly pyriform or obovoid, 3.5–4 × 1 cm.

Type.—*Gentry 10330*, San Ignacio, Baja California, 3 Apr. 1951, "northern slope with volcanic rocks; Cardon-Fouquieria-Jatropha, etc.," deposited in US. Duplicates in DES, MEXU.

Planta parva surculoso-caespitosa paucifoliata; foliis plerumque 15–30 cm longis 2.5–7 cm latis ad medium, crassis lanceolatis conduplicatis glauco-pallidis, margine valde crenata, dentibus supra mammas plerumque 3–8 mm longis flexuosis pallidogriseis debilibus; spina terminali 2–4 cm longa subulata leviter flexuosa pallido-castanea; inflorescentia paniculata cum scapo 2–3 m alta gracili omnino cerasea, ferenti 8–10 ramos parvos remotos; floribus pallido-luteis 44–55 mm longis; ovario cum apice sulcato 23–30 mm longo; tubo vadoso 3–6 mm longo expanso; segmentis 17–22 mm longis 6–7 mm latis aequalibus, pallido-luteis crassis oblongis apice obtusis; filamentis 33–40 mm longis ad basim segmentorum insertis; antheris 18–21 mm longis luteis; capsulis 3.5–4 cm longis 1 cm latis pyriformibus vel obovatis.

Agave cerulata subcerulata strongly resembles *A. subsimplex* across the Gulf along the Sonoran coast, but the relationship is clearly with *A. cerulata*. The depauperate, gray to yellowish rosettes growing on the gypsiferous island, San Marcos, may reach only 10 cm in stature on open exposures. On northward slopes, the rosettes are larger and the panicles reach 2.5–3 m in height with 5 to 8 lateral branches (Fig. 25). Johnston (1924:998) made a similar observation. The ovary and neck of the San Marcos Island flowers collected are more slender than the peninsular ones, but otherwise the flowers are similar, including the waxy bloom on ovaries and pedicels. The short, thick, relatively broad leaves with long teeth on prominent teats provide characters for recognizing this subspecies both in the native habitats and in botanical garden collections. The distribution as far as known is shown in Fig. 21, while habit and habitat appear in Figures 23 (*lower*) and 25.

Agave margaritae

(Figure 32)

Agave margaritae Brandegee. Proc. Calif. Acad. Sci. ser. 2. 2:206. 1889.
Agave connochaetodon Trel. Missouri Bot. Gard. Rep. 22:58. 1912.

Small, compact, cespitose, glaucous gray to yellowish green rosettes with 40–50 leaves; leaves 25 × 10 cm to 12 × 7 cm, ovate to broadly lanceolate, short acuminate, thick, fleshy rigid, concave above, narrowed above the base, the margin crenate with moderate to prominent teats; teeth at mid-blade 4–5 mm or to 8–15 mm long, variously curved or flexed, 1–1.5 cm apart, weakly attached, reddish brown to grayish; spine 2–3 cm long, subulate, shortly and shallowly grooved above, shortly decurrent; panicles 2–3.5 m tall, slender, with 6 to 12 short lateral branches in upper ⅓ of shaft; flowers light yellow, 45–50 mm long; ovary 25–30 mm long, fusiform; tube ca. 10 mm deep; tepals 15 × 5 mm, attenuate; filaments 25 mm long, adnate to the base of tepals (Brandegee) or inserted in the throat of tube (Trelease); capsules 30–50 × 15–20 mm, not stipitate, oblong or pyriform; "seeds 3–4 mm in diameter, smooth" (Brandegee).

Type.—Brandegee s.n., 14 Jan. 1889, Magdalena Island, Baja California, UC. The type consists of two leaves, flowers, and a section of a lateral branch with capsules.

Trelease separated his species *A. connochaetodon* from *A. margaritae* on the basis of longer, more flexuous teeth on larger teats, as he wrote, "with considerable hesitancy that, because of the very different arming of the specimens collected, a second species is recognized for these islands." The population on Isla Margarita shows considerable variation in the size, curvature, teating, and the color of the teeth. Trelease's segregate, in the writer's opinion, is only one of the variants of *A. margaritae* and is accordingly treated as a synonym here. The deep flower tube, the short broad leaves, together with other leaf characters described above distinguish this species from all other peninsular taxa. The relatively deep tube of *A. margaritae* indicates relationship with *A. vizcainoensis*, but on leaf characters they appear specifically distinct, as discussed in the account of *A. vizcainoensis*. *Agave margaritae* is known with certainty only from Magdalena Island, just off the outer coast in the southern part of the peninsula.

Agave sobria

Agave sobria Bdge. ssp. **sobria**

(Figures 26–28; Tables 2, 5)

Agave sobria Brandegee. Proc. Calif. Acad. Sci. ser. 2. 2:207. 1889.
Agave affinis Trel. Missouri Bot. Gard. Rep. 22:56. 1912.
Agave carminis Trel. Missouri Bot. Gard. Rep. 22:55. 1912.
Agave slevinii I. M. Johnston. Proc. Calif. Acad. Sci. ser. 4. 12:1000. 1924.

Small- to medium-sized, usually cespitose, open, few-leaved, bright glaucous-gray rosettes with short stem, or appearing stemless, 50–150 cm in diameter; leaves linear to lanceolate, long-

FIGURE 25. *Agave cerulata subcerulata* on the Island of San Marcos, May 1952.

acuminate, variable, but mostly 80 × 10 cm to 45 × 5 cm, frequently cross-zoned, straight to curved, sometimes twisted, plane to somewhat concave above, thick and concave below towards base, the margin undulate to mammillate; teeth remote, mostly 5–10 mm long with broad, flattened, gray bases, reddish towards apex, variously flexed or straight, mostly 3–4 cm apart; spine acicular, mostly 3–6 cm long, narrowly grooved above; panicles slender, sometimes arching, 2.5–[illegible] m tall with 12–20 short lateral branches, the umbels of flowers compact, nearly globose; peduncular bracts small, triangular, chartaceous; flowers pale yellow, 45–55 mm long, slender; ovary 25–35 mm long, tapering at base, the neck short, scarcely constricted; tepals

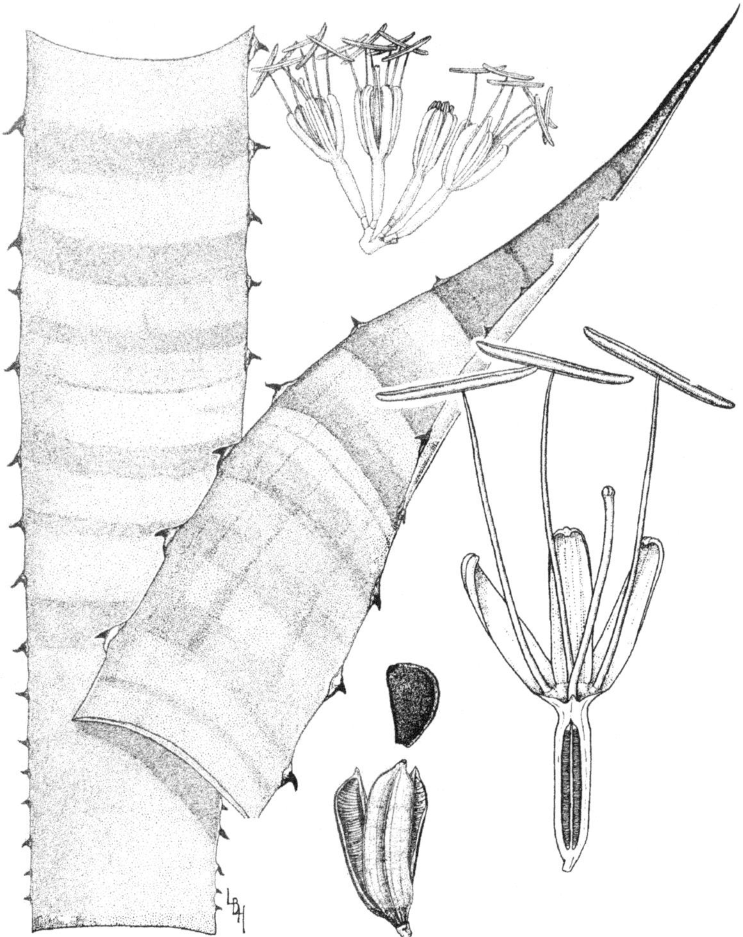

FIGURE 26. *Agave sobria sobria*, drawn from topotypic *Barclay & Arguelles 1989*; leaf, flower cluster and capsules ×½, flower section ×1.3, seed ×2.

about equal, 17–22 × 3–4 mm, very narrow, linear, rather thin, involuting at anthesis, rounded at tip, cucullate, the inner keeled; tube short, spreading, 3–4 mm long, 9–11 mm wide; filaments inserted at base of tepals on nectary disk, 35–47 mm long, slender; anthers 18–23 mm long, centrically attached; capsules 5–6.5 × 1.5–1.8 cm, oblong, thick-walled, short stipitate, apiculate; seeds 7–8 × 5–6 mm, lunate, the margins narrowly winged.

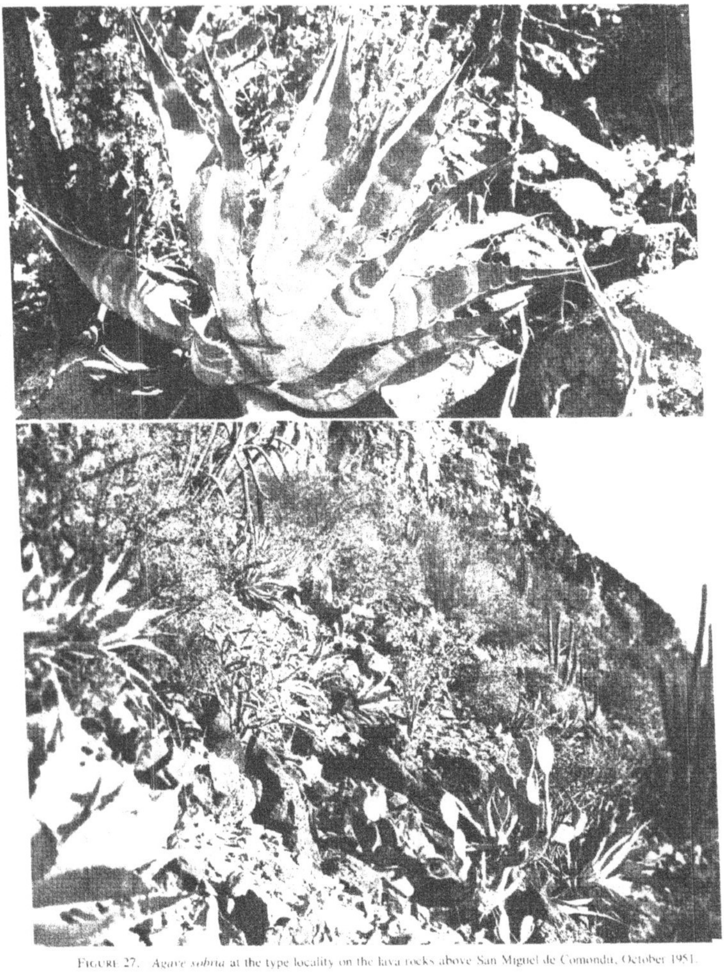

FIGURE 27. *Agave sobria* at the type locality on the lava rocks above San Miguel de Comondú, October 1951.

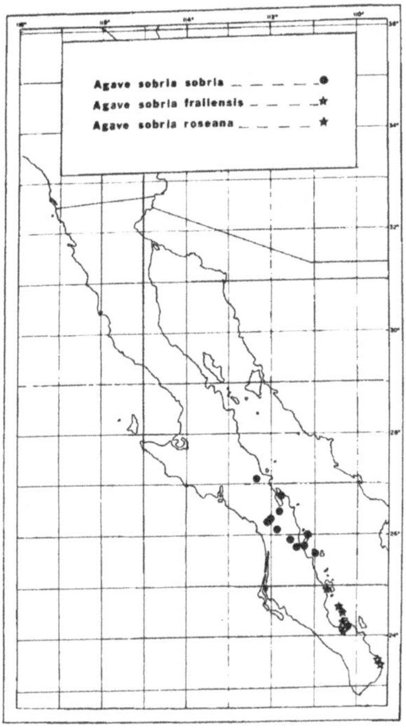

FIGURE 28. Distribution of *Agave sobria* and subspecies, based on herbarium specimens (see Appendix—Exsiccatae).

Type.—*Brandegee 2*, 28 Mar. 1889, Comondú mesas (Fig. 27), Baja California, UC, DS.

Agave sobria is distinguished by its slender flowers with long narrow tepals and its very light-glaucous, long, lanceolate leaves with remote teeth. It is widely scattered but common on both sides of the Sierra de la Giganta mountain axis, from near sea level to 3,500 ft (1,070 m) elevation (Fig. 28). It appears to make its best growth on shadier or northern slopes in canyons, such as below the volcanic rim rock on the talus slope above Comondú. Figure 27 shows the plants at home at the type locality. As examined there in 1951, the rosettes commonly throw out one or more branches from above the base, thus forming cespitose plants. When small, these shoots are readily lifted out of the dry trunk tissue as separate plants with young hard roots started downward. Large rosettes develop when the plants are single or only double. The plants are scattered and do not form dense colonies. On the eastern foot of the Sierra de la Giganta the plants are generally smaller and drought depauperate.

Edibility and Use.—The people at Comondú call this plant "mescal pardo" or "pardito," saying also that it is good for making mescal. This is contrary to what Brandegee (1889) has reported and contrary to his use of the name *sobria* (for sober, or not given to alcohol); so I questioned the inhabitants closely about its edibility and suitability for mescal. They were unanimous in pronouncing it good and that *Agave aurea*, growing in the vicinity, was the one not suitable for mescal. Hence, I believe that Brandegee confused these two agaves on this point in his notes or memory.

However, the Comondú informants, consisting of several older men, said that the agave best for eating and for distilling is called "lechuguilla" and grows in the Sierra de la Giganta. A guide was appointed among them, and I dispatched with him Juan Arguelles, my field assistant. They went by horse and returned with several plants (*Gentry & Arguelles 10324, 10327*) collected on the highland near the base of the mountains above the Llano San Julio. The plants were not flowering, but leaf specimens were pressed, young plants kept for planting, and samples sent into the USDA laboratory in Philadelphia for analysis. With this local information and our present series of specimens, it is now possible to orient the detailed description of agave uses made by Miguel del Barco in the 1770's. He was located for about 30 years (1735–1768) at the Mission of San Javier in the Sierra de la Giganta, and his remarks apparently apply mainly to *A. sobria* and *A. gigantensis*, both of which grow around the mission and were used extensively by the Cochimi Indians. There was a kind that was not eaten, wrote Barco, probably what is now called *Agave aurea*. Most of his remarks I have translated and quoted in the Introduction.

Chemistry.—Table 5 summarizes the sapogenin content of *Agave sobria* as determined by the United States Department of Agriculture on their cortisone-source survey during the 1950's (Wall 1954; Wall et al. 1954a, 1954b, 1955, 1957). The four genins—hecogenin, tigogenin, mano-

TABLE 5. SAPOGENIN CONTENT IN *Agave sobria*. Collated from Wall (1954) and Wall et al. (1954a, 1954b, 1955, 1957). Sapogenins are given in percentages on dry-weight basis: hec. = hecogenin; git. = gitogenin; man. = manogenin; tig. = tigogenin; dead = leaf of dead plant.

				Sapogenins				
Coll. no.	Baja California source locality	Month coll.	Plant part	% total	% hec.	% git.	% man.	% tig.
	Agave sobria sobria							
11291	Comondú	May	leaf	1.6	55			45
11291	Comondú	May	fruit	0.6	100			
11291	Comondú	Nov.	leaf	0.5	X		X	X
11291	Comondú	Nov.	leaf (dead)	0.4	X	X	X	
11811	Sierra de Palmas	May	leaf	1.3			100	
12384	Sierra de la Giganta	Dec.	leaf	0.0				
12384	Sierra de la Giganta	Dec.	leaf	0.6	X		X	X
12387	Bahía Concepción	Dec.	leaf	0.2	100			
12387	Bahía Concepción	Dec.	stem	0.0				
12387	Bahía Concepción	Dec.	leaf	0.2	X		X	X
	Agave sobria roseana							
11274	Islota Gallo	Oct.	leaf	0.7	40	15	10	35
11274	Islota Gallo	Oct.	leaf	0.2	60			40
11277	Espíritu Santo Is.	Oct.	leaf	0.6	57			42
11869	La Paz	May	leaf	0.45	25	10		65
	Agave sobria frailensis							
11264	Punta Frailes	May	leaf	1.1	20	10		70
11264	Punta Frailes	Nov.	leaf	1.4	X			X
11257	Punta Frailes	Oct.	leaf	1.7	10			90
11257	Punta Frailes	Nov.	leaf (dead)	1.3	X	X		X
11858	Punta Frailes	May	leaf	2.2				100
12367	Punta Frailes	Nov.	leaf	1.7	X			X
12368	Punta Frailes	Nov.	stem	0.0				
12369	Punta Frailes	Nov.	leaf (dead)	1.3	X	X		X

genin, and gitogenin—continue as in the *cerulata* complex to chemically characterize the Deserticolae, both in quantity and in random distribution. However, manogenin is lacking in the subspecies *frailensis*, nearly so in ssp. *roseana*, but frequent in ssp. *sobria*. Sapogenin content is conspicuously higher in *frailensis*, averaging 1.5% in leaves, compared to 0.6% in *sobria* and 0.5% in *roseana*. Early summer shows a higher average content of sapogenins in leaves (May 1.3%) than for winter (Nov.–Dec. 0.7%), but the data are scarce and not always strictly comparable. This is the reverse of what was obtained for *A. cerulata*, which is more in the winter-rainfall region, while *A. sobria frailensis* is in the summer-rainfall region. Is higher genin content correlated with dry seasons rather than with temperature? The lack of sapogenin in the edible stems or "cabezas" confirms the statement of the Comondú people that *Agave sobria* is a good source for mescal.

Agave sobria, although it reproduces asexually, is a freely seeding sexual species. It shows a lot of variability among individuals in certain localities, especially in the southern part of its area and on adjacent islands, where isolation has had time to foster genetic divergence. Trelease's *A. roseana*, which Johnston reduced to a variety, appears to represent a natural segregate of *sobria*, and accords with my use of subspecies in this treatise. I am separating it and another geographic variant, hitherto unrecognized, according to the following key and descriptions.

Key to the Subspecies of *A. sobria*

1a. Leaves linear lanceolate, mostly 50–80 cm long, 6–10 times longer than wide, the margin merely undulate; tepals very narrow, 3–4 mm wide .... *sobria*

1b. Leaves broadly lanceolate, 25–50 cm long, 4–5 times longer than wide, the margins prominently mammillate; tepals 4–6 mm wide 2

2a. Rosettes few-leaved, openly spreading, yellow-green; leaves larger, 40–50 cm long, with large teeth few and remote on mammillate margins. Isla Espíritu Santo and adjacent peninsula .... *roseana*

2b. Rosettes with more leaves, compact, urceolate, bluish-gray glaucous; leaves smaller, 25–30 cm long, with smaller teeth closely spaced along mammillate margin. Punta Frailes and vicinity, Cape District .... *frailensis*

Agave sobria Bdge. ssp. **roseana** (Trel.), stat. nov.

(Figures 28, 29, 30 (*right* and *upper*); Tables 2, 5)

Agave roseana Trel. Missouri Bot. Gard. Rep. 22:59. 1912.
Agave sobria var. *roseana* I. M. Johnston. Proc. Calif. Acad. Sci. ser. 4. 12:1002. 1924.

Small, open, few-leaved, cespitose or single, yellow-green rosettes; leaves 35–50 × 7–10 cm, broadly lanceolate, acuminate, concave above, frequently twisted, the margin prominently mammillate, teats 1–1.5 mm high, with remote large flexuous teeth, sometimes bicuspid, the larger 10–25 mm long; spine 5–7 cm long, sinuous to contorted, acicular, narrowly grooved above, castaneous to gray; panicles 2.5–3.5 m tall, slender, with 8 to 12 compact, globose umbels of light yellow flowers in upper third of shaft; peduncular bracts small, triangular; flowers 45–65 mm long; ovary 30–40 mm long, cylindric; tube 4–5 mm long, spreading; tepals 20–22 × 4–5 mm, long-linear, erect-ascending, incurved at apex and cucullate, narrowing with anthesis, inner strongly keeled; tepal sinuses at first closed, wilting open; filaments 35–42 mm long, inserted at base of tepals on rim of tube; stamens 18–20 mm long; capsules 4–4.5 × 1.5–1.8 cm, oblong, short-stipitate.

Type.—Rose 16854, 18 Apr. 1911, Espíritu Santo Island, Baja California, US.

Agave sobria roseana is distinguished by its open, few-leaved rosettes, its short broad leaves with prominent mammillate margins with large irregularly flexed teeth, and its restriction to the southeastern portion of the *sobria* area. It owes its distinction apparently to long isolation on Espíritu Santo Island and the adjacent islets, Islote Gallo and Islote Gallina. It also occurs near La Paz on the peninsula along with plants that are typically *sobria*, but I have had no trouble in separating the few specimens collected there. I did not find evidence of crossing between them, but such may occur. *A. sobria roseana* appears morphologically consistent both on the islands and about La Paz. Variations noted were differences in size of plants and in the size and number of teeth. The population on Islote Gallina showed cespitose plants as common, but all plants on Islote Gallo were singles. Islote Gallo had a rather uniform and dense colony, while elsewhere the plants were widely scattered. As an ornamental this plant is striking because of its bizarre teeth on big teats.

Agave sobria ssp. **frailensis** Gentry, ssp. nov.

(Figures 28, 29, 30 (*lower left*); Tables 2, 5)

Small, compact, sometimes urceolate, glaucous green to bluish glaucous, sparingly cespitose rosettes; leaves mostly 20–35 × 6–8 cm, conduplicate, lanceolate, acuminate, markedly narrowed below, the margin mammillate and with numerous, flexuous, castaneous to graying teeth mostly 6–10 mm long; spine subulate 3–4 cm long, frequently sinuous or contorted, decurrent to upper teeth; panicle slender, 3–4 m tall, with deltoid chartaceous peduncular bracts and 10–15 compact yellow umbels in upper ⅓ to ¼ of shaft; flowers 45–63 mm long, slender; ovary 25–40 mm long, cylindric, with short unconstricted neck; tube 2.5–4 mm deep, small, sulcate from tepal sinuses; tepals 17–23 × 4–6 mm, linear-lanceolate, incurved and cucullate at apex, the outer dark-tinged at tip, the inner with prominent keel; filaments slender 30–40 mm long inserted on rim of tube; anthers 18–22 mm long, centric or excentric; capsules short oblong, 3–4 cm × 1.5–1.7 cm, sessile to short-stipitate, rounded at apex.

Type.—Gentry & Cech 11264, 4 miles [ca. 6 km] N of Punta Frailes, Baja California, 7 Oct. 1951, on granitic slopes, US. Transplant flowered in Murrieta, California, July 1962.

Planta parva surculoso-caespitosa; species foliorum glaucedine cyaneis insignis; foliis plerumque 20–35 cm longis 6–8 latis ad medium, rigescentibus conduplicatis lanceolatis, margine valde crenata, dentibus supra mammas plerum-

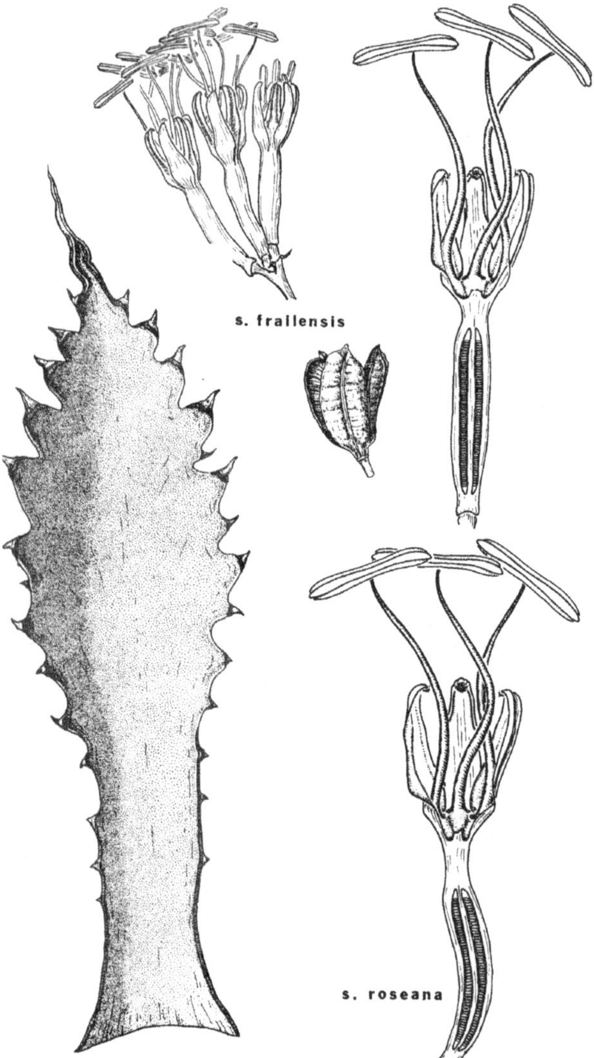

FIGURE 29. *Agave sobria* ssp. *frailensis*: from type, *Gentry & Cech 11264*, leaf, flower cluster, capsule ×½, flower section ×1. *Agave sobria* ssp. *roseana* from topotypic *Gentry & Cech 11277*, flower section ×1.

FIGURE 30. (*Lower left*) *Agave sobria* ssp. *frailensis* near Punta Frailes. (*Right* and *Upper*) *Agave sobria* ssp. *roseana* with inflorescence on the hills east of La Paz. The small, rounded, dense umbels of light-yellow flowers are characteristic.

que 6–10 mm longis castaneis tandem pallido-cinerascentibus; spina terminali aciculata 3–4 cm longa flexuosa castanea vel canescenti decurrenti supra brevi-canaliculata; inflorescentia paniculata gracili, cum scapo 3–4 m alta, ferenti 10–15 ramos congestos; floribus 45–63 mm longis pallido-luteis; ovario cylindrico 25–40 mm longo; tubo brevi 2.5–4 mm longo; segmentis 17–23 mm longis lineari-lanceolatis apice incurvatis cucullatis; filamentis gracilibus 30–40 mm longis ad tubi orificium insertis; antheris 18–22 mm longis luteis; capsulis 3–4 cm longis 1.5–1.7 cm latis breviter oblongis sessilibus vel brevi-stipitatis apice rotundatis.

Agave sobria frailensis differs from its near relative *A. sobria roseana* in its smaller size, more numerous leaves, lighter gray-glaucous color, softer flesh of the leaves, and more numerous, more regular, and smaller teeth. In 1951

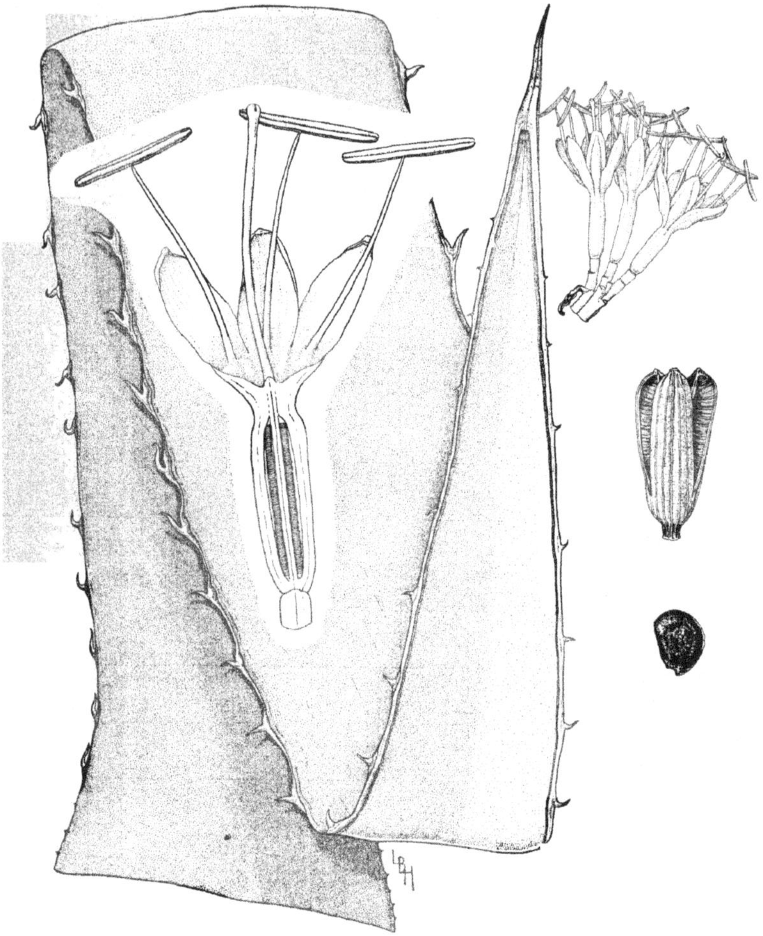

FIGURE 31. *Agave moranii*, drawn from type, *Gentry & McGill 23287*; leaf, flower cluster, capsule ×½, flower section, slightly enlarged.

and 1952 it was found in typical form scattered upon the granitic coastal cerros 2 to 8 miles (3–13 km) northwest of Punta Frailes in the Cape District on the peninsula. This is the only area in which it was observed.

Because of its compact urceolate rosette form and the glaucous, sometimes bluish, color of the leaves, it makes an attractive ornamental. A transplant (*Gentry & Cech 11264*) flowered in Murrieta after 11 years, but numerous offsets

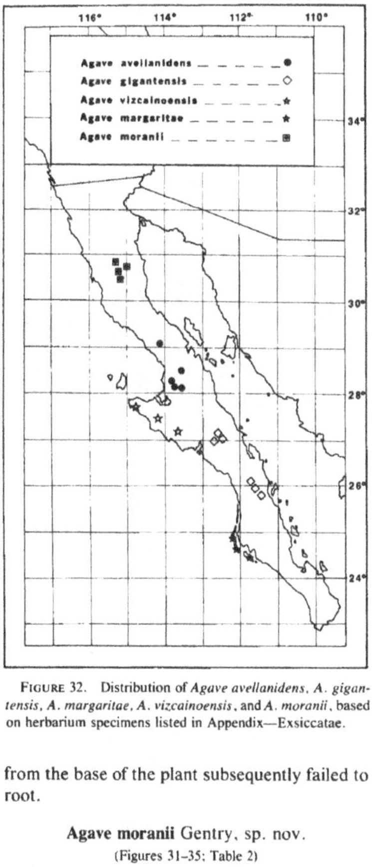

FIGURE 32. Distribution of *Agave avellanidens*, *A. gigantensis*, *A. margaritae*, *A. vizcainoensis*, and *A. moranii*, based on herbarium specimens listed in Appendix—Exsiccatae.

from the base of the plant subsequently failed to root.

Agave moranii Gentry, sp. nov.
(Figures 31–35; Table 2)

Large, single, short-stemmed, mostly light-green rosettes with symmetrical sword-leaved habit, 1–1.5 m tall, 2 m broad; leaves triangular-long-lanceolate, 70–120 × 8–12 cm, deeply guttered, straightly ascending to spreading, rigid, deeply rounded beneath, light green to yellowish green, sometimes glaucous, strongly armed; spine stout 4–6 cm long, nearly white with castaneous tip, broadly grooved above, decurrent to mid-blade; teeth sinuously flexed or curved, flattened, 6–12 mm long through the mid-blade and below, 2–4 cm apart, reduced and more remote towards apex on the white corneous leaf margin, light gray, the base broad or continuous with the margin; panicle 4–5 m tall, stout, frequently swollen and closely bracteate below base of panicle, with 20–30 closely spaced, compact, large, umbellate branches subtended by prominent, lanceolate, reflexing, strong bracts; flowers 50–70 mm long, bright yellow, on slender bracteolate pedicels 1–3 cm long; ovary 25–40 mm long, fusiform, tapered at base and with short, grooved, thick neck; tube shallow, 4–6 × 12–13 mm, discoid-spreading, the nectary ridged by decurrent filaments; tepals 18–24 × 6–7 mm, erect to ascending, broadly linear, acute, with papillate tufted beak, the inner with prominent bulbous keel and 2 costae within, slightly lobed at base; filaments very slender, 35–46 mm long, inserted on rim of nectary and partly on base of tepals; anthers centrically affixed; capsules 5–7 × 1.6–2 cm, long-oblong, stipitate, short-beaked, yellowish to brown; seeds 7–8 × 5–6 mm, lacriform, shiny, the marginal wing smooth, variable in width.

Type.—*Gentry & McGill 23287*, 2–3 miles [3–5 km] SE of Rancho Agua Caliente on E bajada of Sierra San Pedro Mártir, Baja California, elev. ca. 1,500 ft [450 m], 13 June 1973 [l, f, cap] US, isotypes in DES, SD.

Planta grandis breviter caulescens, rosula patulus simplex 1–1.5 m alta 2 m lata; foliis 70–120 cm longis 8–12 cm latis ad medium, anguste trianguli-lanceolatis conduplicatis rigentibus viridibus vel pallidi-viridibus; margine fere recto corneo; spinis marginalibus variabilibus flexuosis ad medium foliorum 6–12 mm longis 2–4 cm separatis, pallido-brunneis vel albis; spina terminali robusta 4–6 cm longa, alba apice castanea supra late canaliculata; inflorescentia paniculata cum scapo 4–5 m alta robusta, ferenti 20–30 ramos approximatos umbelliformes congestos; bracteis grandis siccis triangularibus ad apice aliquando congestis; floribus clarolutеis 50–70 mm longis; ovario fusiformi 25–40 mm longo ad basim angusto; tubo brevi 4–6 mm longo 12–13 mm lato; segmentis 18–24 mm longis 6–7 mm latis, linearibus acutis apice papillatis cucullatis; filamentis gracilibus 35–46 mm longis ad basim segmentorium insertis; antheris 17–21 mm longis luteis ad medium affixis; capsulis 5–7 cm longis 1.6–2 cm latis, oblongis stipitatis brevirostratis; seminibus 7–8 mm longis 5–6 mm latis lacrimiformibus nigris.

FIGURE 33. *Agave moranii* in fine array on northeastern bajada of Sierra San Pedro Mártir, near Rancho Parras, June 1973. Photograph by John McClure.

Agave moranii is distinguished from all other Deserticolae by its large single rosettes of large, long, rigid leaves with white corneous margins apically, by its large-bracteate, stout peduncles, sometimes swollen below the panicles, and by its relatively congested panicles. The only agave with which it may be easily confused, especially from leaf specimens alone, is *A. deserti pringlei* in the same region. As discussed in the description of *A. deserti pringlei*, these two taxa appear to hybridize; without complete specimens they may not be satisfactorily identified. This is the case with *Moran 15256* from Rancho San Pedro Mártir on the western slope of Sierra San Pedro Mártir. A photograph of the plant, however, shows that it is probably surculose—the large panicle, small-bracteate with well-spaced laterals, leading me to place it with *A. deserti pringlei*. The respective habitats appear distinct: *A. moranii* is desertic, while *A. deserti pringlei* is at home in the chaparral zone on the Pacific slopes of the mountains. *Agave moranii* occupies a small area of the southern Sierra San Pedro Mártir and its eastern bajadas, between elevations of 1,500 and 5,000 ft (450 and 1,850 m). The specific name is in honor of Reid Moran, the first botanist to make recognizable collections of the species.

Agave moranii appears aligned phylogenetically with *A. avellanidens* and *A. gigantensis*, all having separate areas along the sierran axis of the peninsula (Fig. 32). As stated above, the three are distinct from other Deserticolae by their non-surculose habits, and their broader and greener leaves.

Near a fork in the road 7 miles (11 km) south of Rancho Agua Caliente near Rancho Parras, there is a fine population of *Agave moranii* (Fig. 33). It is most concentrated on low ridges running out from the sierra side, but it is also scattered on sandy slopes and in the arroyos. Leaf color varies from a light green or glaucous green to yellow-green; a few are lightly zoned. Leaf form (Fig. 33) and armature are fairly constant, but there is some variation in teeth. All healthy plants have outstanding radiating leaf attitudes. Leaf margin is white with a dark brown border along the green. Some peduncles are distinctive with congestion of bracts below the lowest laterals where the peduncle is perceptibly swollen

FIGURE 34. Leaves of *Agave moranii*. Photograph by John McClure.

(Fig. 34). This suggests the same flowering habit as *A. macroculmis*, *A. murpheyi*, and *A. parrasana*, which have the peduncle initiated in the fall, arrested over winter, and a second elongation with the warmer days of spring.

The people living at Agua Caliente have no name to distinguish *A. moranii* from *A. deserti*, which is common in the same vicinity. They stated that the flowers were good to eat. The leaf has an abundance of strong coarse fibers and probably was employed by the Cocopah Indians who formerly inhabited the region. Moran

FIGURE 35. A flowering plant of *Agave moranii* and detail of a thickened peduncle with congested bracts. Photograph by John McClure.

found a single outpost plant of this species 10 miles (16 km) or so towards the east, *Moran 18295*, "only one seen, SE side of San Felipe Valley (Sierra Santa Rosa?)."

Agave avellanidens

(Figures 32, 36, 37; Table 2)

Agave avellanidens TREL. Missouri Bot. Gard. Rep. 22:60. 1912.

Single, medium to large, multi-leaved, green rosettes, 6–12 dm tall, 10–15 dm broad, with stems to 5 dm long; leaves 40–70 × 9–14 cm, broadly linear-lanceolate to ovate, little or unnarrowed base, short-acuminate, thick-fleshy, rigid, smooth, green, the margin straight or undulate and frequently corneous; teeth rather regularly spaced, mostly 1–3 cm apart, but variable in size and curvature, 5–15 mm long, straight or

FIGURE 36. *Agave avellanidens* west of Calmallí, April 1972. The deep narrow panicles and small bracts distinguish it from *Agave shawii goldmaniana*. Photograph by Bruce Gentry.

variously flexed, flattened, dusky gray over brown; spine strong, conical, 2.5–4.5 cm long, broadly grooved above, brown to grayish, strongly decurrent as a corneous margin; panicle 4–6 m tall with 25–35 lateral branches of dense, large, globose umbels of small flowers; flowers pale yellow, drying orange-yellow, 40–70 mm long, slender; ovary 20–40 mm long, fusiform, tapering to base, the neck sometimes constricted, pale yellow; tube 4–6 mm deep, 13–15 mm wide, wide-spreading, ridged within and grooved without from overlapping tepal sinuses; tepals ca. equal, 16–24 mm long, linear-lanceolate, obtuse, the inner with thick keel and sometimes lobate within an inlapping base; filaments slender, 30–45 mm long, inserted on rim of tube; stamens 15–22 mm long, yellow, centric; capsules dark brown, broadly oblong, 20 × 35 mm, not stipitate and scarcely beaked; seeds unknown.

Type.—Brandegee 6, Paraíso [S of San Borja], Baja California, 1 May 1889, UC [1, f, cap].

Agave avellanidens is a *shawii*-like member of the Deserticolae. The elongated many-leaved green rosettes look like *A. shawii goldmaniana* where the two meet and mingle southeast of Punta Prieta. This relationship has been discussed under *goldmaniana*, group Umbelliflorae. When in flower, *A. avellanidens* is distinguishable by the long, narrow, small-bracted panicles of small flowers with short open tubes. These inflorescence characters align it with the Deserticolae rather than with Umbelliflorae, regardless of the shawiian appearance.

The distribution of *A. avellanidens* is poorly known. So far as known, it is limited to a small area in mid-peninsula between latitudes 29° and 28° N. The type locality, Rancho Paraíso, is several miles south of the Misión San Borja in the mountains and apparently has not been visited by any botanist except Brandegee, who was following mule trails. The species is abundant along the old auto road from Calmallí to Rancho Mesquitál (Fig. 36). *Agave shawii goldmaniana* is scattered abundantly along this old road north of Rancho Mesquitál to Punta Prieta, and occasional elongated panicles appear to represent *A. avellanidens*. Certain sightings were also made along the road from San Borja to Rancho Rosalito. Atypical, small-bracted panicles on single rosettes of *A. shawii goldmaniana*, common in the region, made reliable sightings difficult south and east of Punta Prieta. *Agave avellanidens* and its associates of *Fouquieria Yucca*, and *Pachycormus* comprise many exoti and beautiful plantscapes. Perhaps *A. avellanidens* is extensive in that large unknown are about and south of the Sierra Calmallí east o the main road.

Agave gigantensis Gentry, sp. nov.
(Figures 32, 38–40; Table 2)

Single, acaulescent, green to glaucous green rather open, few-leaved, medium-sized rosette 5–10 dm tall, 8–12 dm broad; leaves 40–75 > 11–16 cm, plane, rigid, thick-fleshy, sclerophyllous, smooth, broadly lanceolate, acuminate widest in the middle, markedly narrowed a base, green to glaucous, turning red to purplisl with flowering, and depressed, the margin undulate to prominently mammillate; teeth large sometimes grotesque, mostly 10–20 mm an even longer, the bases thick, frequently 2-3-cuspidate and confluent along upper leaf margins brown to light grayish, variously flexed an curved, generally remote, up to 6–8 cm apart spine 3–6 cm long, strongly subulate, deeply sulcate above, straight or sinuous, gray, long de current as a heavy corneous margin, sometime to mid-leaf; inflorescence 4–5 m tall, slender, th peduncular bracts narrowly lanceolate, thickl chartaceous; panicles rather narrow, in upper ½ to ¼ of shaft, with 15 to 25 rather small umbel of bright pale-yellow flowers; buds waxy white flower 48–60 mm long, slender, with spreadin tepals; ovary slender, fusiform, with short, cor stricted, slightly grooved neck; tube short dis coid, spreading, bulbous with tepal bases, 4– mm deep, 11–13 mm broad; tepals 18–25 × 5– mm, linear, rounded and briefly hooded at apex the inner with prominent keel, grooved withi and strongly overlapped at base by outer; fila ments 30–45 mm long, variable in length, slender, inserted on rim of tube with base of tepals anthers 18–25 mm long, somewhat excentric capsules 3.5–4 × 1.2–1.5 cm, oblong, non-stip itate, non-beaked; seeds unknown.

Type.—Gentry & McGill 23320, above Rancho San Sebatián, Sierra de las Palmas, 31 miles [ca. 50 km] by road W San Bruno, Baja California, elev. 3,800–5,000 ft [1,150–1,5 m], 20 June 1973, US. Isotypes DES, SD, MEXU.

Planta simplex nonsurculosa, rosula 5–10 dı alta, 8–12 dm diametro lata glaucifolia vel vi

FIGURE 37. *Agave avellanidens: upper*, a series of "normal" leaves; *lower*, a form with bizarre armature of 2- to 3-cuspid teeth. Photographed west of Calmallí.

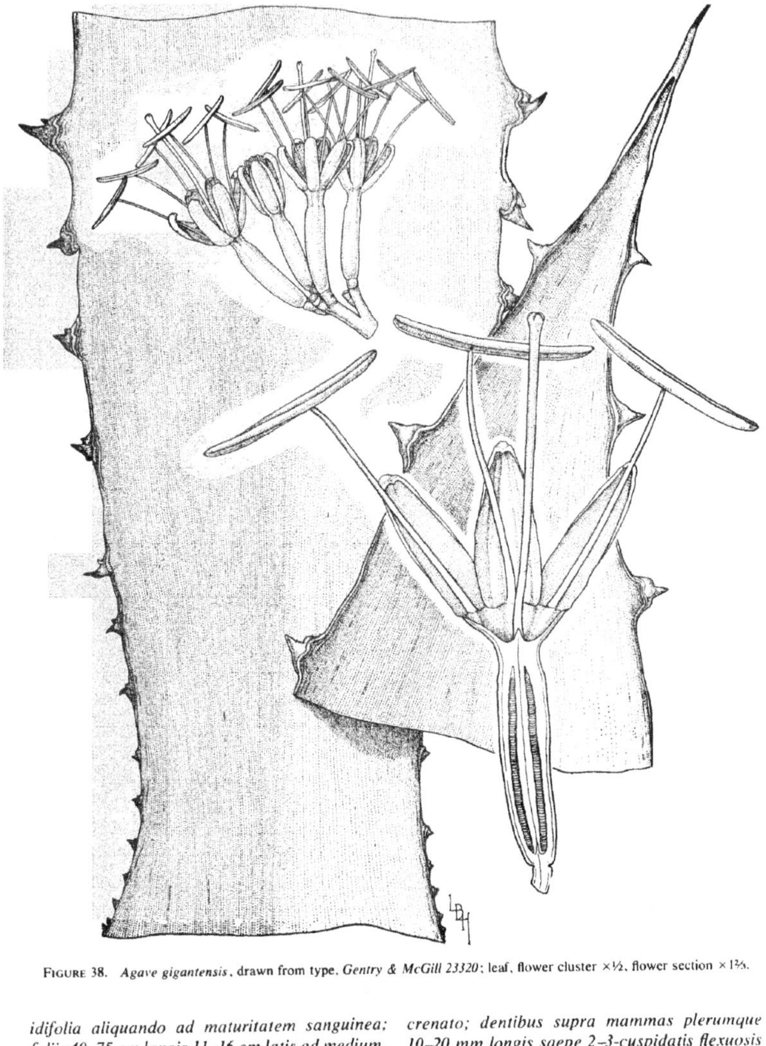

FIGURE 38. *Agave gigantensis*, drawn from type, *Gentry & McGill 23320*; leaf, flower cluster ×½, flower section ×1⅔.

idifolia aliquando ad maturitatem sanguinea; foliis 40–75 cm longis 11–16 cm latis ad medium, late lanceolatis planis rigidulis crassi-succulentis laevibus, ad basim angustatis; margine valde crenato; dentibus supra mammas plerumque 10–20 mm longis saepe 2–3-cuspidatis flexuosis vel curvatis, remotis usque ad 6–8 cm separatis; spina terminali 3–6 cm longa robusta subulata

FIGURE [illegible]. *Agave gigantensis*. (*Upper*) a transplant flowering at Murrieta, California, with geotropic flowers and a hummingbird. (*Lower*) A mature rosette on Sierra de las Palmas.

recta vel sinuosa albocastanea, supra profunde canaliculata; inflorescentia paniculata cum scapo 4–5 m alta gracili ferenti 15–25 ramos umbelliformes congestos parvos; bracteis anguste triangulis crassi-chartaceis; floribus pallide luteis fere albis 48–62 mm longis; gemmis albis; ovario fusiformi 25–35 mm longo apice constricto; tubo vadoso 4–5 mm longo 11–13 mm diametro; segmentis 18–25 longis 5–6 mm latis ad medium linearibus, apice brevi-cucullatis; filamentis 30–45 mm longis variabilibus gracilibus ad apicem tubi insertis; antheris 18–25 mm longis luteis excentricis; capsulis 3.5–4 cm longis 1.2–1.5 cm latis, oblongis non stipitatis; seminibus ignotis.

The small waxy pale flowers with short spreading tube in a slender panicle places *A. gigantensis* in the Deserticolae. As with other members of this alliance, the habit and leaves provide specific distinction. The leaves are remarkable in their broad form above a narrowed base, their singular quality of color, bud printing, and firm sclerophyllous smoothness, and bizarre variable large teeth on mammillate margins (Figs. 38, 39). The nearest phyletic relatives appear to be the disjunct *A. avellanidens* and *A. moranii* to the north. The name *gigantensis* is given because the distribution is coincident with the Sierra de la Giganta range of the southern peninsula (Fig. 32).

On Sierra de las Palmas (also known locally as Sierra Campana), *Agave gigantensis* is closely associated with an oak woodland community that mantles the craggy tops of these volcanic mountains (Fig. 40). *Nolina beldingii* is a conspicuous member of this community. Southward, *A. gigantensis* has been observed and collected on the brushy slopes west of the Cerro Giganta itself and along the scarp-rim country above Loreto (see Exsiccatae citations). Altogether, it is known to range from elevations between 2,000 and 5,000 ft (600 and 1,520 m) between latitudes 27° N and 25° N. This is a small area, and *A. gigantensis* shapes up as another rare endemic plant. Fortunately it occupies a very rugged and largely inaccessible habitat, which in this age of destructive man will favor its survival. The rancher at San Sebastián, who guided us upon the mountain, stated that this agave is more abundant on "Sierra del Potrero," northward in this same mountain range. It is a symmetrical, beautiful plant that could well b valued and cultivated by the sophisticated agav fanciers. It does not reproduce vegetatively however, and seeds would be needed to keep i generating.

The name "lechuguilla" is applied to thi agave by the people of Comondú. It is consid ered as one of the best for distilling into mescal Of the five samples of leaves sent in for analysis only two showed sapogenin in minor amounts 0.3% hecogenin. Undoubtedly, this is one of th more edible agaves to which Miguel del Barc referred in his early account (quoted in the In troduction).

Agave vizcainoensis Gentry, sp. nov.

(Figures 32, 41; Table 2)

Plants acaulescent, small to medium, surcu lose or single, the rosettes 30–50 cm tall, 50–9 cm wide, few-leaved and open in habit; leave lanceolate, 25–40 × 6–10 cm, glaucous gray t green, sometimes reddish, broadest in middle narrowed above base, thick-fleshy, rather rigid the margin undulate, corneous above with de current spine; teeth largest along mid-blade, 5 10 mm long, 1–3 cm apart, slender or broadl flattened, dark brown to grayish, nearly straigh or curved, spine 2.5–4 cm long, stoutly subulate mostly rather straight, brown to grayish, shal lowly grooved above, long decurrent; panicle 2 3 m tall with 8 to 15 spreading lateral umbellat branches in upper half of shaft; flowers yello 65–75 mm long, on stout small-bracteolate ped icels 8–12 mm long; ovary green, 36–41 mr long, in fresh flower neck scarcely constricte but drying very narrow, 6–8 mm long; tube fun nelform, 8–12 mm deep, 15 mm broad, bulgin at filament insertion; tepals 21–26 × 4–5 mm linear-lanceolate, involute above spreadin base, obtuse, cucullate, strictly erect at anthe sis; filaments 55–70 mm long, inserted somewha unequally 7–9 mm above base of tube; stamen 21–26 mm long, excentric; pistil to 70–85 mr long with broad lobate stigma; capsules 5–7 × 2 cm, oblong, stipitate, rostrate, striate, brown seeds 6–7 × 4.5 mm, lacrimiform, black, mar ginal wing narrow.

Type.—Gentry 7469, Cerro Tordillo and vicinity, Sierra Vi caíno, Baja California, 12–13 March 1947; elev. 400–800 [120–250 m]; desert of dispersed succulent trees and suffrute cent shrubs, UC; isotypes, DS, ARIZ, MEXU, DES.

FIGURE 40. *Agave gigantensis* in native habitat on Sierra de las Palmas. *Nolina beldingii* in background. Photographed by John McClure, June 1973.

Planta acaulis simplex vel surculosa grisea, 3–5 dm alta 5–9 dm lata, paucifoliata; foliis lanceolatis 25–40 cm longis 6–10 cm latis, pallidis vel griseo-viridibus carnosis; margine undulata, dentibus plerumque 5–10 mm longis 1–3 vel 4 cm separatis, surrectis vel flexuosis gracilibus aliquando complanatis, atrobrunneis vel cinerascentibus; spina terminali 2.5–5 cm longa robusta recta vel flexuosa, brunnea vel cinerascenti, supra canaliculata longe decurrenti; inflorescentia paniculata 2–3 m alta diffusa, ferenti 8–15 ramos laterales; floribus luteis 65–80

FIGURE 41. Type collection of *Agave vizcainoensis*, *Gentry 7469*, an isotype sheet in the Gentry Herbarium.

mm longis; ovario viridiflavo; tubo infundibuliformi 8–12 mm longo, apice 15 mm diametro; segmentis 21–26 mm longis 5–6 mm latis lineari-lanceolatis, involutis cucullatis, rectis sub anthesi; filamentis 55–70 mm longis gracilibus, 7–9 mm supra basim tubi insertis, antheris 21–26 mm longis, excentricis; pistillo usque ad 70–85 mm longo stigmate trilobato; capsulis 5–7 cm longis 2 cm latis oblongis stipitatis rostratis brunneis; seminibus 6–7 mm longis, 4.5 mm latis lacrimiformibus nigris, margine in alam anguste expansa.

The relatively deep flower tubes of this species suggests affinity with *A. margaritae* of the

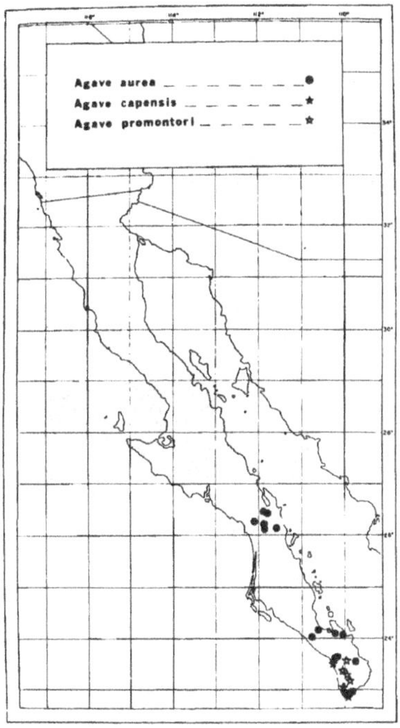

FIGURE 42. Distribution of the group Campaniflorae, based on herbarium specimens listed in Appendix—Exsiccatae.

same geographic region, but the latter, with its short, broad, long-toothed leaves and small size, appears distinct. Neither species is well known; the collections are few and incomplete. The species do not show close relation with other members of the Deserticolae, but they are assigned here for want of a more suitable section to receive them. Their small size, surculose habit, shape of leaf and armature, and the small scarious bracts of the peduncle exclude them from the *A. shawii* complex. The deepening tube of subspecies *simplex* in the pivotal complex of *A. deserti*, at almost opposite ends of the Deserticolae area, seems also to permit their inclusion in the Deserticolae.

The plants growing about the Picachos de Santa Clara are more robust and less surculose than those in the Sierra Vizcaíno proper and resemble *A. gigantensis*. *Gentry 7713* agrees well with the type in leaf morphology, but the tepals are relatively long in relation to the shorter tube (Table 2). Another specimen (*Gentry 7693*), also from the Picachos de Santa Clara, is more extreme with tepals to 35 mm long.

Group Campaniflorae

Trelease, Missouri Bot. Gard. Rep. 22:44. 1912.

Large to small, perennial or multiannual plants with rather open, short-stemmed rosettes; leaves green, rather soft succulent, long-lanceolate, frequently sigmoid towards apex, with regular, close-set, moderate teeth on noncorneous margins. Panicles large, diffuse, in upper half of shaft subtended by small scarious bracts; flowers campanulate with deep, ample tubes, deep-set filaments, red to purple on outside, yellow within; capsules short-oblong to ovoid, rather thin walled. Southern half of peninsular Baja California.

The leaf surface in the Campaniflorae is moderately roughened but noticeably thickened. The stomates are slightly depressed and without rims and with overlapping polar lips in the montane *A. promontorii*. The number of stomata per mm^2 ranges from 18 to 31 on the upper leaf surface, a relatively low density, setting them apart from the Deserticolae and Umbelliflorae. Further particulars of epidermal structure are given in the companion study of Gentry and Sauck (1978).

The Campaniflorae occupy one of the most isolated areas of the continental Agavaceae (Fig. 42). The pre-Cretaceous granitic pediment on which they grow has long been isolated from the contemporary formation of the Sierra Madrean mainland. This isolation seems to be reflected by the Campaniflorae in their consistent conformation to the distinct campanulate flowers with broad tubes and thin-margined tepals, all with a certain succulent texture difficult to describe. The regular, moderate, close-set teeth on the soft succulent leaves are also consistently maintained through the group. The flower proportions appear unusually variable (Fig. 43 and Table 6). I have no explanation for this. They do not appear useful for species separation. The connective gland with the filament-anther fixation in *A. capensis* has not been observed elsewhere (Fig. 45). The habit of axillary budding with resultant large clusters of many rosettes,

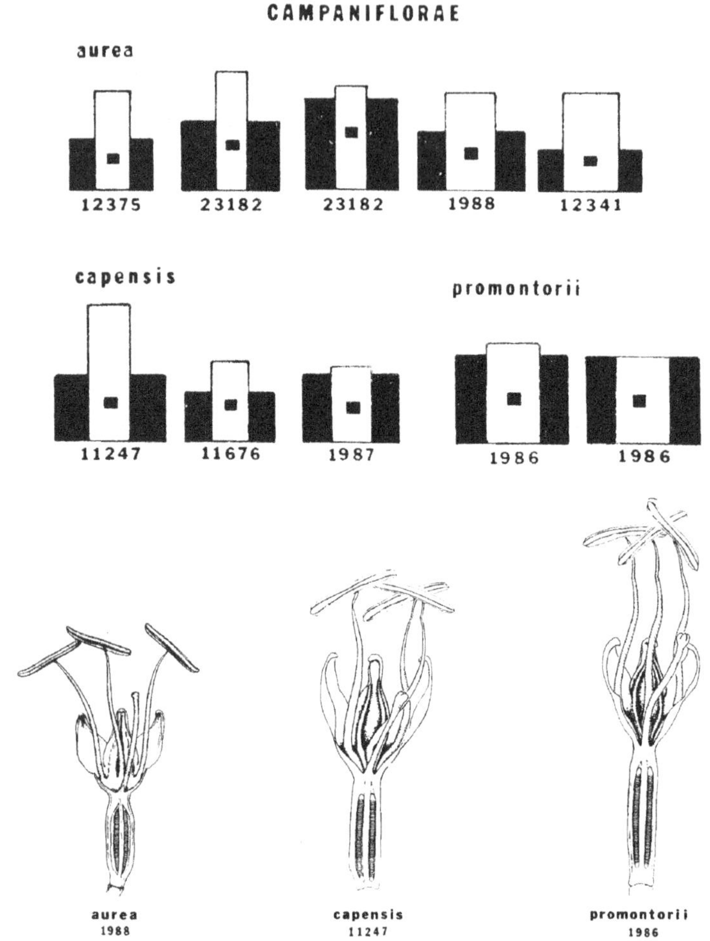

FIGURE 43. Floral ideographs of the group Campaniflorae, showing relative proportions of the tube (black) to outer tepal (white column), and level of inseertion of filament (black square). Measurements are listed in Table 6.

found in *A. capensis*, is regarded as a relic character, widely scattered in the genus from the Umbelliflorae of the northwest to the Striatae and Polycarpae of central and eastern Mexico.

The rainfall silhouettes in Figure 44 show that the Campaniflorae occupy moister habitats than other agave groups of the Peninsula. Rain falls predominantly in summer-fall from storms of tropical origin. The large growth-form of *Agave promontorii* appears to reflect the higher rainfall.

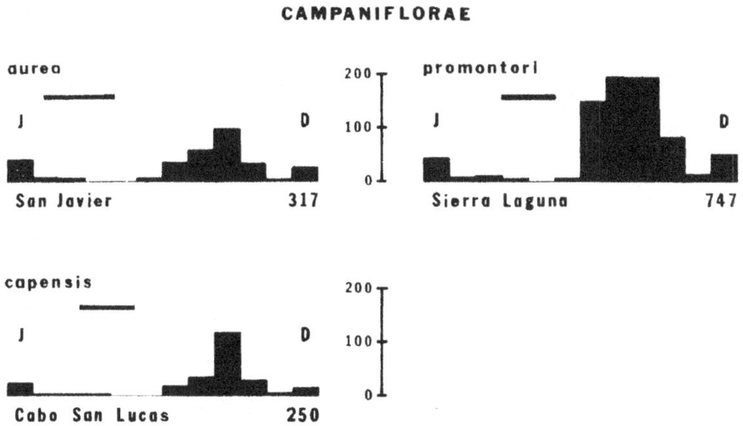

FIGURE 44. Rainfall (silhouettes) and flowering (bars) perimeters of the Campaniflorae. Relevant meteorological stations with mean annual rainfall in millimeters. Data from Hastings 1964. Flowering periods based on herbarium specimens and field observations, supplemented by plants in cultivation.

It is notable also that flowering occurs in the dry spring months (Fig. 44), when birds or other pollinators would be, by thirst, more attracted to the liquid flower nectar. The Campaniflorae are free-seeding outbreeders, not given to vegetative reproduction.

The prehistoric uses of the Campaniflorae plants are unknown. *Agave aurea* alone is known to have been exploited for fiber for a short period late in the 19th century, and a large population on the lavas about Comondú was probably decimated. The following tables (7 through 9) of chemical analyses show that some steroids are common in all three species, with hecogenin being the most common. As with other peninsular species, however, the incidence of the sapogenins appears fugitive within species, by seasons, and by physiological plant stages. Until further investigations can show how three percent (3%) or more sapogenin content can be consistently produced for harvest, the cultivation of *Agave* for sapogenin alone is not economically feasible.

Key to Species of the Campaniflorae

1a. Rosettes large, nonsurculose multiannuals; leaves 60–150 cm long; panicles broad; filaments without apical gland. ... 2

1b. Rosettes smaller, perennials by axillary budding; leaves mostly 35–60 cm long; panicles narrow; filaments with an apical gland at anther fixation. Cape District .... *A. capensis* (p. 72)

2a. Rosettes generally 0.7–1 m tall; leaves smaller, more pliant, 7–12 cm broad. Sierra de la Giganta and Cape District .... *A. aurea* (p. 78)

2b. Rosettes 1.5–2.3 m tall; leaves larger, thicker, relatively rigid, 11–17 cm broad. Sierra Laguna, Cape District .. .. *A. promontorii* (p. 81)

Agave capensis Gentry, sp. nov.

(Figures 42–47; Tables 6–7)

Agave brandegeei in hort. & herb.

Plants perennial by axillary budding, eventuating in large clusters with short-stemmed, small, open rosettes, 6–8 dm tall, 8–12 dm broad; leaves mostly 30–60 × 4–7 cm, narrowly lanceolate, straight to arching, commonly sigmoid towards apex, concave above, convex below, light glaucous green, soft, brittle, succulent, with undulate noncorneous margin; teeth regular, 4–5 mm long, mostly 1–2 cm apart, mildly curved, on short mammillate bases, reddish brown to grayish; spine 1.5–3 cm long,

TABLE 6. FLOWER MEASUREMENTS (IN MM) OF THE GROUP CAMPANIFLORAE. (*—measurements from dried flowers; **—measurements from fresh or pickled flowers.)

Taxon & locality	Ovary	Tube	Tepals	Fil. insert. & length		Anther	Total length	Coll. no.
aurea								
Comondú	32	8 × 14	16 × 6	5	17	18	55**	12375
	29	8 × 14	16 × 6	6	35	19	54**	12375
	20	9 × 14	16 × 6	5	38	20	45**	12375
	30	11 × 17	19 × 5.5	7–8	44	23	67**	23182
	34	11 × 16	19 × 5	8–7	45	23	63**	23182
	36	15 × 15	16 × 5	10–9	46	20	66**	23182
	39	14 × 15	17 × 5	9–8	45	21	70**	23182
Todos Santos	30	9 × 18	17 × 9	7 & 5	41	21	55**	1988
	30	10 × 18	15 × 9	7 & 5	40	20	54**	1988
	20	7 × 8	17 × 10	6–5	45	19	43**	12341
	21	6 × 18	15 × 10	5–4	43	19	43**	12341
Cape Region	32	7 × 8	14 × 7	4–5	40	19	52*	Bdge.
capensis								
Cabo San Lucas	34	12 × 20	20 × 7	6	38	28	65**	11247
	31	10 × 20	23 × 7	7	43	27	63**	11247
	27	8 × 13	13 × 6	6–7	26	15	48**	19676
	24	8 × 14	13 × 6	5–6	27	16	44**	19676
	25	11 × 16	12 × 7	5–6	32	19	48**	1987
	28	11 × 16	13 × 7	5–6	40	20	52**	1987
promontorii								
Sierra Laguna	42	14 × 20	16 × 9	8 & 6	54	26	72**	1986
	36	14 × 20	14 × 9	8 & 6	50	24	64**	1986
	41	14 × 20	15 × 9	8 & 6	50	24	70**	1986

subulate, dark brown, short decurrent for 1–2 cm; panicles mostly 2.5–3.5 m tall with 15–24 lateral branches in upper ⅔ to ½ of shaft, the laterals up to 30 cm long, ascending; umbels small; bracteoles very small, 2–3 mm long, narrow-triangulate; flowers in bud reddish brown or purplish, opening yellow inside, 50–65 mm long; ovary green, 25–35 mm long, thick, 3-angled, furrowed in unconstricted neck; tube 8–14 mm deep, 15–20 mm broad, 6-bulged, 6-furrowed from tepal sinuses; tepals equal, 13–23 × 6–7 mm, lanceolate, incurved and rather thin; filaments 30–43 mm long, yellow, inserted 5–7 mm above base of tube, the apex swollen with colorless glands; anthers 15–27 mm long, excentric, yellow or bronze; capsules 25–35 × 15–17 mm, ovoid, scarcely stipitate, apiculate; thin walled; seeds not seen.

Type.—Gentry & Fox 11247, Cabo San Lucas & vicinity, Baja California. Young plant collected 5 Oct. 1951; flowered in Murrieta, California, July 1964; deposited in US.

Planta caespitosa perennis per gemmas rosulatas axillares; foliis angustis plerumque 30–60 cm longis 4–7 cm latis linearibus vel lanceolatis canaliculatis arcuatis vel sigmoideis fragilibus molliter succulentis glauco-viridibus, margine undulatis; dentibus ad medium foliorum 4–5 mm longis plerumque separatis 1–2 cm, paulum curvatis castaneis vel griseis; spina terminali 1.5–3 cm longa subulata atrobrunnea, decurrenti usque ad 1–2 cm; inflorescentia paniculata cum scapo 2.5–3.5 m alta, ferenti 15–24 ramos umbelliformes parvos; floribus 50–65 mm longis in alabastro rubris vel purpureis sub anthesi ferrugineis intra luteis; ovario 25–35 mm longo viridi rotundato-triquetro; tubo 8–14 mm longo 15–20 mm diametro, campanulato 6-sulcato ferruginescenti; segmentis 13–23 mm longis 6–7 latis infra medium, lanceolatis cucullatis tenuibus incurvatis aequalibus; filamentis luteis 30–43 mm longis 5–7 mm supra tubi basim insertis, glanduloso-tumidis ad apicem; antheris 15–27 mm longis luteis vel ferrugineis excentricis; capsulis 2.5–3.5 cm longis 1.5–1.7 cm latis, ovoideis subsessilibus apiculatis valvis tenuibus; seminibus ignotis.

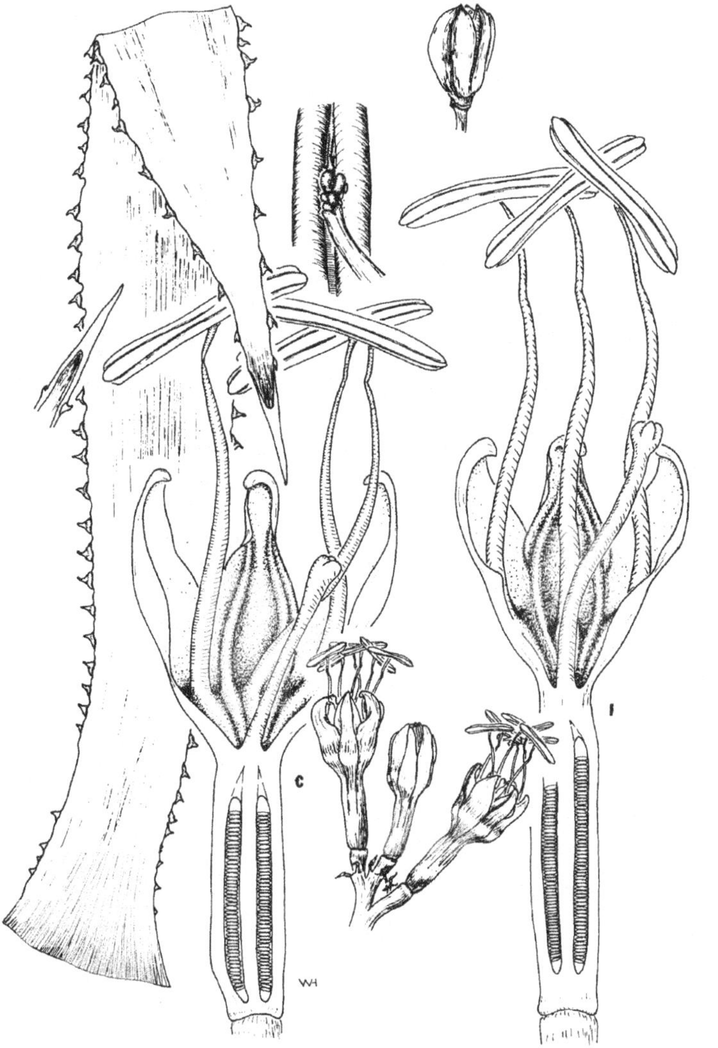

FIGURE 45. (*C*). *Agave capensis* drawn from type: leaf, flower cluster, capsule ×½, flower section ×2⅓, apex of filament enlarged to show gland. (*P*) *A. promontorii*, flower section ×2⅓.

FIGURE 46. *Agave capensis*, a large axillary-branching clone in Huntington Botanical Gardens, California, January 1951.

Agave capensis appears to be an islandic endemic of the Cape District. It grows with the sparse open shrubbery on the arid, neutral, well-aerated soils on granitic slopes from near sea level to 1,000 ft (ca. 300 m) or more above sea level. Scattered single rosettes northwest of Punta Frailes appeared similar to those of *A. capensis*, but their immaturity at time of sighting leaves their identity doubtful. This is also true of a collection (*Gentry 11301*) in the shady canyon above La Purísima, which probably represents a narrow-leaved shade form of *A. aurea*.

The flower structure of *Agave capensis* is very similar to that of its sectional relatives, *A. aurea* and *A. promontorii*, but it is distinguished by its small narrow leaves and clustered growth habit. Figure 46 shows a large clump in Huntington Botanical Gardens in 1951, which Mr. Hertrich, the superintendent, stated was about 40 years old. The photographs in Figure 47, taken in the same year at San José del Cabo, show both single and branching rosettes and are both very similar to the Huntington clone. A specimen transplanted from San José del Cabo in 1951 flowered at Murrieta in July 1964 and is selected as the type. Subsequent to the July flowering, some of the axillary rosettes also flowered, but none of the flowers at Murrieta developed fertile fruits.

The connective gland or swelling found on the stamen where the anther is affixed to the filament (Fig. 45) is a structure peculiar to the Campaniflorae. While well developed in *A. capensis*, it was also noted in minor form on a putative hybrid, × *A. aurea*. These two species appear to be close, even genetically compatible, as discussed under *A. aurea*, but the origin of the filament gland appears to lie with *A. capensis*.

FIGURE 47. *Agave capensis* at Cabo San Lucas; (*upper*) a cluster of rosettes showing axillary branches, and (*lower*) a single rosette.

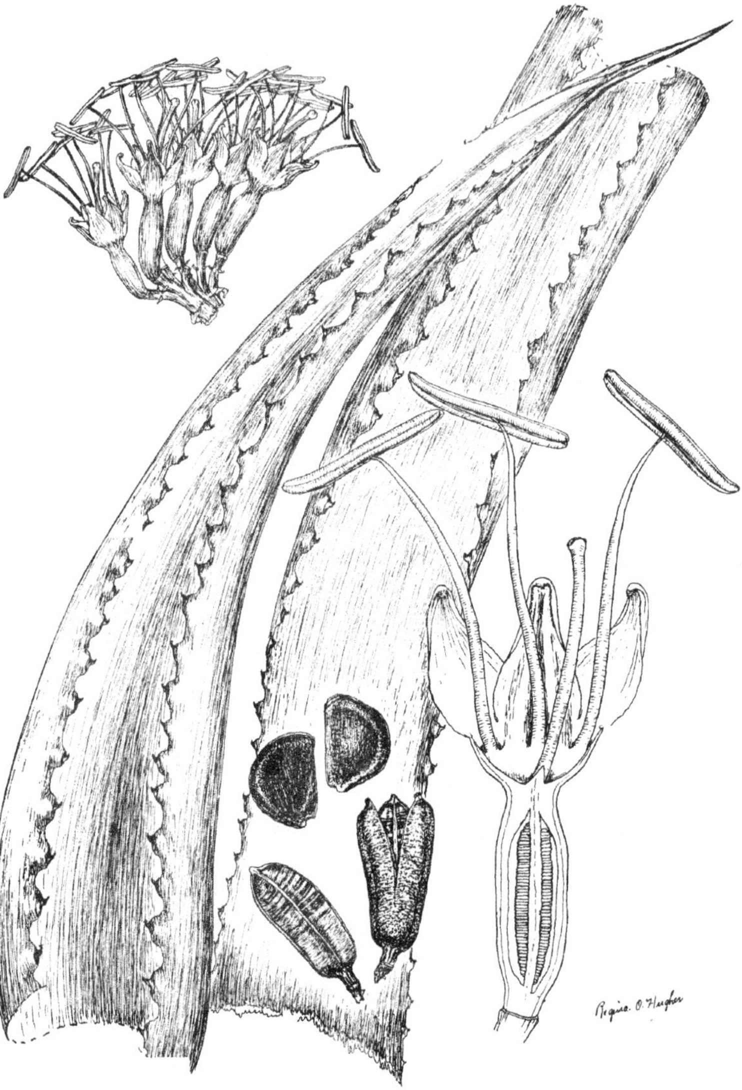

FIGURE 48. *Agave aurea*, leaf from *Gentry 11295*, flower cluster from *Barclay & Arguelles 1988*, capsules ×½, flower section ×1½. Leaf is incurled from drought.

In the Huntington Botanical Gardens in San Marino, and in other California gardens, this species has been known as *Agave brandegeei* Trel. I had hoped to maintain this name, but examination of the type specimen, "Cape Region mountains, Brandegee Sept. 20, 1899," in the herbarium of the Missouri Botanical Garden shows that this name must be abandoned. The type specimen consists of a partial leaf and flowers of *A. aurea* Brandegee and flowers of *A. sobria* Brandegee. Furthermore, in describing the species, Trelease excluded flowers of *A. aurea* and described those of *A. sobria* to accompany the leaf specimen. The two parts of the type, therefore, are synonymized under *A. aurea* and *A. sobria*. The Cape species must be recognized and described as a new species with a new name. It is a curious coincidence that both the Brandegee and Purpus collections from the Cape region are of the same mixture. Obviously, there was some mishandling of specimens after the collections were made, but by whom and when is unknown.

Agave aurea

(Figures 42–44, 48–49; Tables 6, 8)

Agave aurea Brandegee, Proc. Calif. Acad. Sci. ser. 2. 2:207. 1889.

Single short-stemmed, rather open, green, graceful rosettes 10–12 dm tall, 15–20 dm broad, with widely arching leaves; leaves 63 × 7 cm to 110 × 12 cm, average 86.3 × 8.6 cm, linear to long-lanceolate, pliant, guttered, rounded below, thickly fleshy towards base, green to somewhat glaucous, the margin straight to undulate; teeth moderate, regular, mostly 4–7 mm long, 1–2 cm apart, straight or moderately curved cusps from low angular bases, dark brown to light brown; spine subulate, 25–35 mm long, dark brown or grayish red, with a short narrow groove above, shortly decurrent or decurrent as a dark corneous margin through 8 to 10 teeth bases; panicles 2.5–5 m tall, broad, the peduncle reddish and with remote, quickly drying lanceolate bracts; umbels broad, congested, 15 to 25 in upper half of shaft, red to purplish in bud and fruit; flowers opening yellow to orange-yellow, 43–70 mm long, campanulate; ovary reddish, 25–35 mm long, slender, angular-cylindric, with constricted furrowed neck 6–10 mm long; tube 8–14 mm deep, 14–18 mm broad, evenly grooved from tepal sinuses; tepals 16–19 mm long, lanceolate, acute, incurving, the outer larger with tufts of white papillae below distinct hood, the inner with long keel and yellow, thin margin involuting at anthesis; filaments stout, 35–45 mm long, those of outer tepals inserted 5–10 mm above base of tube, slightly higher than those of inner tepals; anthers 19–23 mm long; capsules oblong, 45 × 17 mm to 35 × 15 mm, non-stipitate, rounded apiculate, reddish, drying light brown; seeds 7–8 × 5–5.5 mm, irregularly lunate, dull black.

Type.—*Brandegee s.n.*, Purísima, Baja California, 13 Feb. 1889, UC. Consists of marginal sections of a large leaf, a small leaf, 5 flowers, pieces of 2 capsules. The plant does not grow in the sandy valley at La Purísima; doubtless collected on the volcanic mesas near that settlement.

Characteristics and Distribution.—*Agave aurea* is the most abundant species in the group Campaniflorae. It is easily recognized by the long narrow lanceolate green leaves arching out to form an open, spreading rosette, by the broad, rather diffuse, reddish panicles, and by the bright yellow flowers from reddish buds and ovaries. The teeth are moderate and regularly spaced (Figs. 48, 49). It forms scattered and sometimes massive populations on the extensive lava fields of the western slopes of the Sierra de la Giganta, mostly between elevations of 1,000 and 3,500 ft (ca. 300 and 1,070 m) above sea level. It is also widely scattered on the granitic lower slopes of the Cape District at lower elevations. It is lacking or rare on the sandy Magdalena plain.

In the latitude of Todos Santos, an *A. aurea* population shows forms similar to both *A. promontorii* and *A. capensis*. Here there are forms like *aurea* in the size of their rosette and in the color and form of their leaves, but with the softer leaves and more proximal teeth of *promontorii*. Here also are small cespitose-rosette forms with softer narrower leaves of low fiber content like those of *capensis* at Cape San Lucas. Large inflorescences with well-developed fruits and ovules are among these forms, indicating interfertility. It appears that *A. aurea* may have invaded the islandic Cape District after it was connected to the peninsula in Pleistocene times and is now receiving genes from the insular *promontorii* and *capensis* populations.

Fiber of A. aurea *and* A. promontorii.—The fiber of *Agave aurea* was at one time harvested in the vicinity of Comondú, where this agave grew in abundance upon the lava beds. Accord-

FIGURE 49. *Agave aurea* on the lava beds north of Comondú, April 1973. Photograph by Bruce Gentry.

ing to Comondú informants, this was towards the end of the 19th century. In 1951 some adobe ruins several kilometers back in the canyon below Comondú marked the site of the decorticating mill. The operation was extensive, and cart trails may still be traced through the lava fields where the leaves were hauled in. Nelson (1921:38) in 1905 noticed these same roads when approaching Comondú by horse from La Purísima. It is not now known just why this operation terminated; was the accessible supply exhausted, or did the operation not profit?

The fiber of *Agave aurea* is fine and of excellent spinning quality, but yield of fiber per leaf

TABLE 7. SAPOGENIN CONTENT IN *Agave capensis*. Collated from Wall (1954) and Wall et al. (1954a, 1954b, 1955, 1957). Sapogenins are given in percentages on dry-weight basis: hec. = hecogenin; git. = gitogenin; man. = manogenin; tig. = tigogenin.

Coll. no.	Baja California source locality	Month coll.	Plant part	Sapogenins % total	% hec.	% git.	% man.	% tig.
11247	Cabo San Lucas	Oct.	leaf	0.4	35			65
11247	Cabo San Lucas	Oct.	stem	0.0				
11247	Cabo San Lucas	May	leaf	1.0	100			
11247	Cabo San Lucas	Nov.	leaf	0.0				
11250	Cabo San Lucas	Oct.	leaf	0.8	80			20
11250	Cabo San Lucas	May	leaf	0.55	20			80
11250	Cabo San Lucas	Nov.	leaf	3.6	X		X	X
11847	Cabo San Lucas	May	fruit	0.3	20		30	50

is low, and compared with henequen or sisal could not be competitive for twine and rope. In 1952 leaf samples of *A. aurea* and *A. promontorii* were sent to the Mexican Fibre Company in Tampico, Tamaulipas, for testing. Grady Venable, in charge of that operation, made the following report.

1) *Agave promontorii*. The fiber is excellent and, it is believed, would be acceptable to the market. The yield of fiber per leaf, however, is too low for commercial operation, but it could probably be made to produce more fiber under proper planting conditions. If the chemical properties are more valuable than the henequen now grown on El Carrizal, Tamaulipas, it might be cultivated commercially.

2) *Agave aurea*. This fiber, while of excellent texture, cannot be produced commercially unless it is found that the yield can be increased. It cannot be commercially produced unless its chemical content should prove exceptionally valuable.

1. *Agave promontorii*:
   - A. 4 leaves—total weight, approx. 2,000 grams.
   - B. Average weight per leaf—500 grams.
   - C. Yield of fiber—total 65 grams.
   - D. Percentage yield per gross—3.25%.
   - E. Yield per leaf—average 16 grams.
   - F. Milling characteristics—good.
   - G. Texture of fiber—fine, softer than henequen.
   - H. Spinning factor—rates as excellent.
   - I. Break test—our estimate 20% weaker than henequen.
   - J. Length recovered fiber—80 cm, good.
   - K. Bleaching—color (sun), good.
   - L. Pulp adherence—minimum, good.

2. *Agave aurea*:
   - A. Approximately as above.
   - B. Approximately as above.
   - C. Yield—approximately 26 grams.
   - D. Percentage yield—1.21%.
   - E. Yield per leaf—average 6 grams.
   - F. Same as above.
   - G. Same as above. Better than *promontorii*.
   - H. Same as above. Better than *promontorii*.
   - I. Same as above.
   - J. Length recovered fiber—60 cm, short commercially.
   - K. Same as above.
   - L. Same as above.

I supplied the test samples and can add some additional qualifying notes. The *A. aurea* samples sent were collected near Todos Santos in the Cape District from plants having atypically little fiber in the leaves (see discussion above). Leaves decorticated by hand from Comondú plants were found to have a stronger and more abundant fiber than the sample sent. Transportation difficulties and working conditions prevented getting Comondú samples to Tampico for analyses.

Chemistry.—The tables (7, 8 and 9) giving the sapogenin contents of the Campaniflorae, enable one to judge further any commercial possibilities. The selection of individual plants with abundant fiber together with high sapogenin content for breeding could materially raise the commercial potential.

TABLE 8. SAPOGENIN CONTENT IN *Agave aurea*. Collated from Wall (1954) and Wall et al. (1954a, 1954b, 1955, 1957). Sapogenins are given in percentages on dry-weight basis: hec. = hecogenin; git. = gitogenin; man. = manogenin; tig. = tigogenin; dead = leaf of dead plant.

Coll. no.	Baja California source locality	Month coll.	Plant part	Sapogenins				
				% total	% hec.	% git.	% man.	% tig.
11198	La Paz	Sep.	leaf	0.7	100			
11253	San José del Cabo	Oct.	leaf	0.5	70			20
11253	San José del Cabo	May	leaf	0.4	100			
11255	Las Cuevas	Oct.	leaf	0.0				
11283	NW of La Paz	Oct.	leaf	0.0				
11295	Comondú	Oct.	leaf	0.1	100			
11295	Comondú	May	leaf	0.0	100			
11295	Comondú	Oct.	leaf (dead)	0.3				
11297	Comondú	Oct.	leaf	0.55	90		10	
11297	Comondú	May	leaf	0.34	87		13	
11297	Comondú	Dec.	leaf	1.4	X	X	X	X
11823	Todos Santos	May	leaf	0.0				
11823	Todos Santos	May	fruit	1.6	45		40	15
11886	Comondú	May	fruit	0.5	5	30		65
12383	San Javier	Dec.	leaf	1.2	100			
12338	La Paz	Nov.	leaf	1.2	X	X	X	X
12339	La Paz	Nov.	leaf	1.1	100			
12342	Todos Santos	Nov.	leaf	0.0				
12341	Todos Santos	Nov.	leaf	0.7	X		X	
12420	Comondú	Dec.	leaf	0.1	X		X	

Agave promontorii

(Figures 42–44, 50; Tables 6, 9)

Agave promontorii TREL. Missouri Bot. Gard. Rep. 22:50. 1912.

Large, single, green, open rosettes, 1–2 m or more tall, 2–2.5 m broad, with thick stems; leaves 10–15 dm long, 11–17 cm wide, lanceolate fleshy succulent, stiff, usually concave above, thick at base, straight to arching, green to soft glaucous green, the margins about straight and closely set with regular straight to curved teeth, mostly 4–8 mm long, 5–10 mm apart, reddish brown; spine 3–5 cm long, conic-subulate, dark brown, short-decurrent, narrowly sulcate above; inflorescence 5–9 m tall, massive, the peduncle reddish and with conspicuous deltoid bracts; panicles large, broad, with 25–30 diffuse umbels on recurving laterals in upper half of shaft, the buds and ovaries red to purplish; flowers short-pedicellate, 60–75 mm long, campanulate; ovary 36–42 mm long, 3-angled-cylindric, scarcely narrowed at base, 6-furrowed and narrowed in neck; tube 14–15 mm long, 20 mm broad, cup-shaped, 6-bulged, with deep furrows between bulges; tepals 14–16 × 9 mm, triangular-lanceolate, equal, thin, reddish purple on outside, yellow within, the outer plane and overlapping inner at base, the inner with yellow margins and low, broad, 4-lined keel, with inner costae pronounced and wide apart; filaments 50–55 mm long, inserted unequally at 8 and 6 mm, oval in cross-section, yellow; anthers 24–26 mm long, yellow; capsules shortly pyriform-oblong, 15–20 × 30–35 mm, rather stipitate, beaked; seeds narrow, 4–5 × 6–9 mm. (Fl. description from *Barclay and Arguelles 1986*; cap. and seed description from Trelease 1912).

Type.—*Nelson & Goldman 7437*, Sierra de la Laguna, Baja California, 21 Jan. 1906, US.

Agave promontorii is a large handsome plant, becoming massive with thick stems and thick juicy leaves in fertile situations (Fig. 50). The arching attitude of the leaves combined with their soft glaucous green and regular close-set teeth make it an attractive ornamental. It is, however, rare in gardens because it does not offset and must be renewed with seedlings. The plants growing in Balboa Park, San Diego, and my plants near Murrieta, California, did not set seed. Night freezes of 24 F (−44 C) injured the leaves causing apical dieback.

The species is known only from the granitic

TABLE 9. SAPOGENIN CONTENT IN *Agave promontorii*. Collated from Wall (1954) and Wall et al. (1954a, 1954b, 1955, 1957). Sapogenins are given in percentages on dry-weight basis: hec. = hecogenin; git. = gitogenin; man. = manogenin; tig. = tigogenin; dead = leaf of dead plant.

				Sapogenins				
Coll. no.	Source locality	Month coll.	Plant part	% total	% hec.	% git.	% man.	% tig.
	BAJA CALIFORNIA							
11218	Sierra Laguna	Oct.	leaf	0.3	55			55
11218	Sierra Laguna	Oct.	stem	0.0				
11229	La Burrera	Oct.	leaf	0.75	90		10	
11229	La Burrera	May	leaf	0.0				
11218	Sierra Laguna	Nov.	leaf	3.4	X		X	X
12346	Sierra Laguna	Nov.	leaf (dead)	0.4			X	X
12347	Sierra Laguna	Nov.	inflo.	0.2	X			X
12353	Sierra Laguna	Nov.	leaf	0.8	X		X	
12349	Sierra Laguna	Nov.	leaf	0.7	X			X
12351	Sierra Laguna	Nov.	leaf (dead)	0.9			X	X
12352	Sierra Laguna	Nov.	leaf	0.9	X		X	
12350	Sierra Laguna	Nov.	leaf	0.5	100			
12356	Sierra Laguna	Nov.	leaf	0.5	X		X	X
12344	Burrera	Nov.	leaf	0.4	100			
12345	Burrera	Nov.	leaf	0.8	X	X	X	X
	CALIFORNIA							
3233	Balboa Park, San Diego	Nov.	leaf	0.0				

mountains of the Cape District. It attains best development in the higher elevations from 3,000 to 6,000 ft (ca. 900–1,800 m) above sea level, coincident with the better moisture conditions about the Sierra de la Laguna. It flourishes, for instance, on the open, precipitous, rocky, southwest slopes above the Rancho Burrera. Although it was observed in the shade of trees, it does not thrive in the crowded chaparral thickets common on the Sierra Laguna. Strays from the montane populations are common along the rocky arroyos issuing from the canyons on the west, as at Rancho La Burrera. Due to its limited and remote range, *A. promontorii* is another rare *Agave* scarcely known to man.

Except for variations in color and teeth, *Agave promontorii* appears well-knit morphologically. The flowers show it to be closely related to the other members of the Campaniflorae. However, its large size in both rosette and leaf distinguish it clearly from both *A. aurea* and the still smaller cespitose *A. capensis*. It was noted earlier that some of the *aurea* population around Todos Santos bears resemblance to *promontorii*, but no individuals of the latter were observed in the vicinity. The apparent introgression there was probably only between *A. aurea* and *A. capensis*, whose populations meet in the low coastal elevations there and near Cabo San Lucas.

I failed to record any local uses of *Agave promontorii*, nor have I seen any account on this subject by other field botanists. The thick, succulent, young, flowering shoots look tempting as a primitive sweet, and quite possibly they were eaten by the original Amerindians, who, because of their susceptibility to European diseases, disappeared from the Cape District soon after Spanish colonization in the 18th century. The leaf fiber is fine, readily twillable, and of good length, strength, and quality, but leaf yield relative to the heavy succulent bulk or weight is low. Analysis of the leaf fiber is given and discussed above under *A. aurea*. The sapogenin content is indicated in Table 9, but here, as elsewhere in the peninsular *Agave*, the percentage content of sapogenin is rather fugitive. The commercial possibilities of the combined fiber and chemical products remain but little investigated.

GROUP UMBELLIFLORAE

TRELEASE, Missouri Bot. Gard. Rep. 22:44. 1912

Small to large, freely seeding, long-generation, and frequently long-stemmed plants com-

FIGURE 50. *Agave promontorii* in Balboa Park, San Diego, California, with large but sterile infructescences

monly branching from leaf axils, eventuating in fragmented supine clones; leaves broad, thickly succulent, stiff, well armed, mostly bright green; inflorescence stout, compact, with large, many-flowered, umbellate branches subtended by large, succulent, sheathing bracts; flowers large, fleshy, with ample tubes and strong divergent stamens; ovaries thick; capsules large, thick walled. Baja California.

The *shawii* complex is distinguished primarily by perennial plants that flower repeatedly, according to the development of axillary branches. Flowering is not at regular seasons, but occurs only as the branching rosettes mature. This appears, from the few records available and from field observations, to require twenty to forty years. Records show that a specimen of ssp. *goldmaniana* (Desert Botanical Garden No. 42)

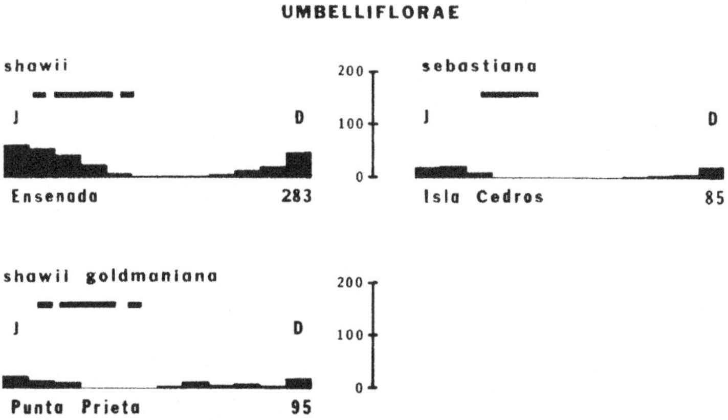

FIGURE 51. Rainfall (silhouettes) and flowering (bars) of the Umbelliflorae. Relevant meteorological stations with average annual rainfall given in millimeters. Data from *Atlas Meteorológico de México* (Servicio Meteorológico Mexicana 1939) and Hastings 1964. Flowering periods based on herbarium specimens and field observations, supplemented by plants in gardens; uncertainty expressed by broken line.

collected as a small plant by George E. Lindsay near San Andrés, Baja California, in the mid-thirties, required more than 35 years to mature and flower in 1972. The trunk at that time was about 2 m long, had declined upon the ground, and is presently being continued by two large branch rosettes. Further exposition of this unique habit is given in the Introduction. The individual rosettes, especially when they have declined upon the ground, rooted, and become separate or independent, may be regarded as multiannuals, a common reproduction habit among agaves. Moran (1964) reported an age of 31 years to maturity for a plant of ssp. *goldmaniana* at the Mina Desengaño in Central Baja California.

It should be noted that this perennial habit is not maintained throughout the various populations. Numerous individuals have been observed that mature and die without leaving any branched rosettes to follow. This is strikingly apparent, for instance, in the *goldmaniana* populations south of Punta Prieta, where single rosettes far outnumber the clustered rosettes (Fig. 58). There are no other consistent morphological features to distinguish these single plants from clustered plants, and they are regarded as intra-subspecific. The causes of this variation in habit, whether environmental or genetic, are not known. It may be regarded as a growth-habit potential that is not always individually fulfilled.

The Umbelliflorae are further distinguished by their green, smooth leaves with relatively thin and unchanneled cuticle, by their rimmed stomata without polar lips, and entire surface-pore margins. The number of stomata on the upper leaf surface ranges from 40 to 47 per mm². Illustrations and further particulars of epidermal structure are given in Gentry and Sauck (1978).

The closest group relation of the Umbelliflorae is not clear. The Campaniflorae of the southern half of the peninsula are well separated by their narrower, nonrigid, fine-fibered leaves with regular teeth, and by their shorter, thinner, campanulate flowers in deep, small-bracteate panicles. One species of the Campaniflorae, *A. capensis*, like *A. shawii*, branches from the leaf axils. But I do not believe this constitutes close affinity because other, more distant relatives in the subgenus *Littaea* also have this growth habit. The Umbelliflorae and Campaniflorae occupy distinct geographic regions with no sympatric members. The only observed possibility of gene interchange by members of Umbelliflorae with other group representatives is that between *A. shawii goldmaniana* and *A. avellanidens* of the Deserticolae.

The Umbelliflorae occupy an arid climate, but

are unique among agave groups in being confined to a Mediterranean type of climate with winter-spring rainfall and dry summers. The rainfall silhouettes of Figure 51 show only 85 & 95 mm (3 & 4 inches) of annual rainfall for *sebastiana* and *goldmaniana*, respectively, yet they are robust plants with broad leaves. However, the rainfall lack is greatly ameliorated by frequent fogs that condense and bathe the leaves, and reduce insolation and temperatures. Even the more inland populations of *goldmaniana* are regularly visited by inland-moving fogs. The fogs increase and dilute the nectar in the cupullate flowers, which is frequently copious in early morning hours.

The flowering seasons are also indicated in Figure 51. The flowering seasons of *A. shawii* and ssp. *goldmaniana* are relatively long and commonly start with the spring rains, but they appear to vary from year to year and have been observed to flower in the fall.

The Deserticolae diverged with the increasing aridity of continental climates, while Umbelliflorae remained in the more equable, conservative, maritime environment. In concert with western mountain forming and diversification of climate into isolated habitats, the Deserticolae evolved into relatively numerous species and subspecies, while in the Umbelliflorae speciation remains indistinct. In our present state of knowledge, it is not possible to state when, in geologic or biologic time, the group divergence began.

The axil-branching growth habit, a type of perennialism among the agaves, is found generally in the more equable tropical climates and appears to be an ancient characteristic. Its disjunct occurrence on the maritime California coast appears as a relic survival. In the Agavaceae, axil-branching is a homologous permagene character.

Agave shawii ssp. *shawii* and *A. sebastiana* look very similar in habit, as they both have elongate, tightly imbricated rosettes, short thick peduncles, and wide congested panicles of very similar, large, succulent, bright-yellow flowers. They are the maritime ecotypes of the species, while *goldmaniana* is the desertic ecotype. Historically, one can speculate that in pre-Pleistocene times, *shawii* and *sebastiana* forerunners were continuous along the coast from San Diego to Cedros Island. With the orogenic changes in the Pleistocene, there was also a drying of cli-

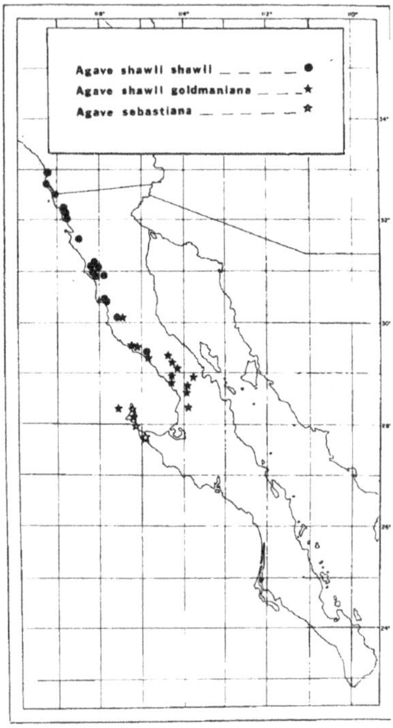

FIGURE 52. Distribution of the Umbelliflorae, based o herbarium specimens listed in Exiccatae (Appendix).

mate. Desert shrub moved over to the coast be low Rosario, and *A. goldmaniana* moved wit the shrub and displaced and separated *A. shawi* into two populations. *A. goldmaniana* becam contaminated with *shawii* genes and remaine or regressed to a subspecies. *A. sebastiana* sur vived as a disjunct on the Cedros Island grou and, as of the present, has reached a specifi level of taxonomic divergence.

During the early 1950's, samples of leaves stems, and inflorescences were collected o *Agave shawii* to determine the sapogenin con tent. The results of this chemical survey of th Agricultural Research Service are brought to gether in Table 10. The sapogenin content wa found insufficient as practicable sources for fab ricating cortisone and other steroidal drugs Other economic aspects are noted below.

TABLE 10. SAPOGENIN CONTENT IN *Agave shawii*. Collated from Wall (1954) and Wall et al. (1954a, 1954b, 1955, 1957). Sapogenins are given in percentages on dry-weight basis: hec. = hecogenin; git. = gitogenin; man. = manogenin; tig. = tigogenin; wh.pl. = whole plant.

				Sapogenins				
Coll. no.	Source locality	Month coll.	Plant part	% total	% hec.	% git.	% man.	% tig.
	Agave shawii shawii BAJA CALIFORNIA							
10282	Km 57 S of Tijuana	Mar.	leaf	0.0				
10283	Km 57 S of Tijuana	Mar.	stem	0.0				
10281	Km 57 S of Tijuana	Mar.	infl.	0.0				
10285	San Telmo	Mar.	leaf	0.38				
10379	Socorro	Apr.	fruit	0.35				
10379	Socorro	Apr.	leaf	0.0				
11180	Socorro	Apr.	infl.	0.56				
11694	Rosario	Apr.	fruit	0.2				97
11698	Rosario	Apr.	leaf	0.25	55		35	10
11699	Rosario	Apr.	leaf	0.15	100			
10383	Rosario	Apr.	fruit	0.0				
	CALIFORNIA							
—	Huntington Bot. Gard.		leaf	0.0				
	Agave shawii goldmaniana BAJA CALIFORNIA							
10349	Marmolito	Apr.	leaf	0.0				
10355	Tinaja Yubay	Apr.	infl.	0.14			100	
10355	Tinaja Yubay	Apr.	leaf	0.06	50		50	
10355	Tinaja Yubay	Apr.	stem	0.1				
10355	Tinaja Yubay	Apr.	infl.	0.15	60	27		
10359	Tinaja Yubay	Apr.	leaf	0.3	63		10	20
11317	Marmolito	Oct.	fruit	0.25	55	28	17	
11318	Punta Prieta	Oct.	leaf	0.3			X	
11319	Punta Prieta	Oct.	leaf	0.0				
11964	Mina Desengano	Oct.	fruit	1.3	70		30	
11936	Mesquitál	Oct.	wh.pl.	0.7	X		X	
12402	Punta Prieta	Dec.	wh.pl.	0.8			X	
12401	Punta Prieta	Dec.	leaf	0.9	X		X	
12400	Punta Prieta	Dec.	leaf	0.6	X			

Key to Taxa of the Umbelliflorae

1a. Panicle about as broad as long, subtended by a cluster of large, succulent, sheathing bracts below lowest laterals; laterals 8 to 15; leaves short, generally only 20–40 cm long, short-acuminate, usually less than 3–3.5 times longer than broad 2

1b. Panicle longer than broad, the bracts below lowest laterals smaller and not closely clustered; laterals mostly 20 to 30; leaves longer, generally 40–60 cm long, long-acuminate. Mid-peninsula .... *shawii goldmaniana* (p. 93)

2a. Panicle with a rounded crown, the subtending bracts large, succulent, red to purple; leaves bright green. Northern peninsular coast .... *shawii shawii* (p. 86)

2b. Panicle with a flat crown, subtending bracts small, scarious, pink to yellow; leaves glaucous gray to green. Cedros Island .... *sebastiana* (p. 96)

Agave shawii

Agave shawii Engelm. spp. **shawii**

(Figures 51–57; Tables 10, 11)

Agave shawii ENGELM. Trans. Acad. Sci. St. Louis 3:314, 370. 1875.

Agave orcuttiana TREL. Missouri Bot. Gard. Rep. 22:47, 1912.

Agave pachyacantha TREL. Ibid. p. 48.

Single or cespitose, compact, green, small to medium rosettes with short to long stems (to 2

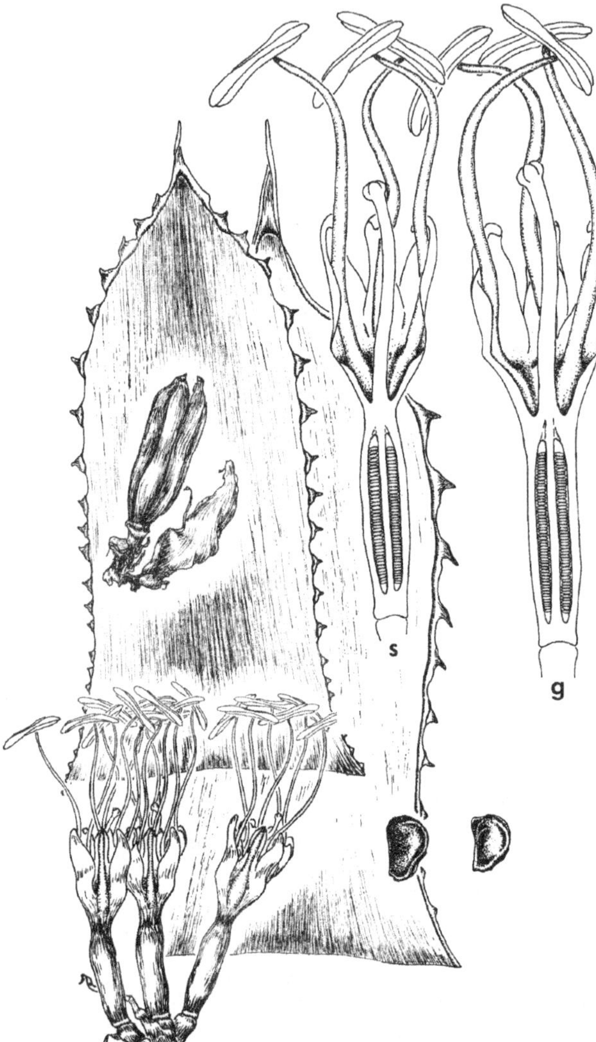

FIGURE 53. (*S*), *Agave shawii* ssp. *shawii*: leaf, flower cluster ×1½, flower section ×1, from *Gentry 23190*; larger leaf from *Gentry 23154*, capsule ×½, seeds ×1½. (*G*), *A. shawii* ssp. *goldmaniana*: flower section ×1 from *Gentry 19974*.

TABLE 11. FLOWER MEASUREMENTS (IN MM) OF THE GROUP UMBELLIFLORAE. (*— measurements from dried flowers; **—measurements from fresh or pickled flowers.)

Taxon & locality	Ovary	Tube	Tepals	Fil. insert. & length		Anther	Total length	Coll. no.
shawii shawii								
La Misión	37	12 × 19	25 × 8	8	73	25	74**	23190
	37	16 × 18	23 × 7	8–7	72	26	77**	23190
	37	14 × 18	24 × 7	9–7	72	25	77**	23190
	42	14 × 18	26 × 6	9–10	68	25	81**	23190
	41	16 × 19	25 × 7	10–9	70	26	82**	23190
Arroyo Seco	38	16 × 16	23 × 5	11–12	55	21	78**	11716
	29	16 × 16	24 × 5	11–12	52	20	67**	11716
Socorro	47	12 × 20	28 × 7	10		30	88**	11714
	56	14 × 20	26 × 7	9	65	30	96**	11714
	39	13 × 20	30 × 5	7–8	53	27	82**	11967
	53	15 × 18	25 × 6	9–10	40	27	92**	19967
	55	13 × 18	27 × 5	9–10	60	26	94**	19967
E of Rosario	53	15 × 20	30 × 7	10	64	28	97**	23154
	55	16 × 21	30 × 7	10–8	70	28	101**	23154
	50	14 × 19	28 × 5	10–8	70	25	92**	23154
	54	14 × 17	25 × 7	10	55	26	92**	19968
	55	13 × 18	27 × 5	10	60	27	95**	19968
shawii goldmaniana								
Punta Prieta	40	13 × 24	25 × 8	9–10	65	26	78**	19974
	51	14 × 24	27 × 9	10–11	65	27	92**	19974
	50	13 × 20	29 × 8	9–10	53	33	92**	19975
	40	9 × 14	25 × 7	8	52	21	76**	19975
	37	11 × 14	18 × 6	8	45	22	67**	19975
Tinaja Yubay	34	9 × 15	22 × 6	7	45	26	65**	Harb.
	35	11 × 15	24 × 6	8	42	26	71**	Harb.
Punta Prieta	50	10 × 21	25 × 8	8–9	47	24	85**	23161
	49	11 × 20	25 × 9	9–10	58	24	84**	23161
	34	9 × 17	21 × 8	7	50	26	65**	1985
	34	9 × 18	22 × 8	6–8	50	25	65**	1985
W of San Borjas	57	16 × 22	24 × 10	12–13	55	24	96**	23166
	41	12 × 23	23 × 9	10–11	70	24	78**	23166
	45	12 × 23	26 × 9	8–9	75	26	82**	23166
	36	10 × 18	24 × 8	8–6	46	23	70**	23166
San Andrés	51	12 × 20	34 × 9	9–10	70	26	96**	42
	50	11 × 20	29 × 8	8–9	70	21	90**	42
	51	10 × 20	32 × 8	7–8	58	27	93**	42
Mesquitál	33	11 × 17	19 × 7	9–10	42	20	63**	11935
	34	10 × 17	21 × 7	8–10	42	20	64**	11935
	29	9 × 17	18 × 7	6–8	45	19	56**	11937
	26	9 × 17	15 × 7	5–7	42	18	50**	11937
sebastiana								
San Benito Is.	38	15 × 18	16 × 5	8–10	42	21	69*	17449
	42	15 × 19	23 × 5	10–11	50	20	80*	17430
	35	20 × 22	20 × 6	14–15	42	21	75*	17431
	44	15 × 20	20 × 6	12	45	20	80*	17431
Natividad Is.	52	14 × 20	21 × 5	8–9	60	18	86*	15142

m), erect to decumbent, frequently branching from leaf axils; leaves mostly 20 × 8–10 cm to 50 × 20 cm, ovate to linear-ovate, short acuminate, glossy light to dark green, cuticle slightly asperous, thickly fleshy, rigid, plane to slightly hollowed above, about as wide in mid-blade as at base; spine 2–4 cm long, acicular, broad at base, openly grooved above, dark reddish brown to gray, straight or sinuous, decurrent for 8–10 cm or along entire leaf as corneous margin;

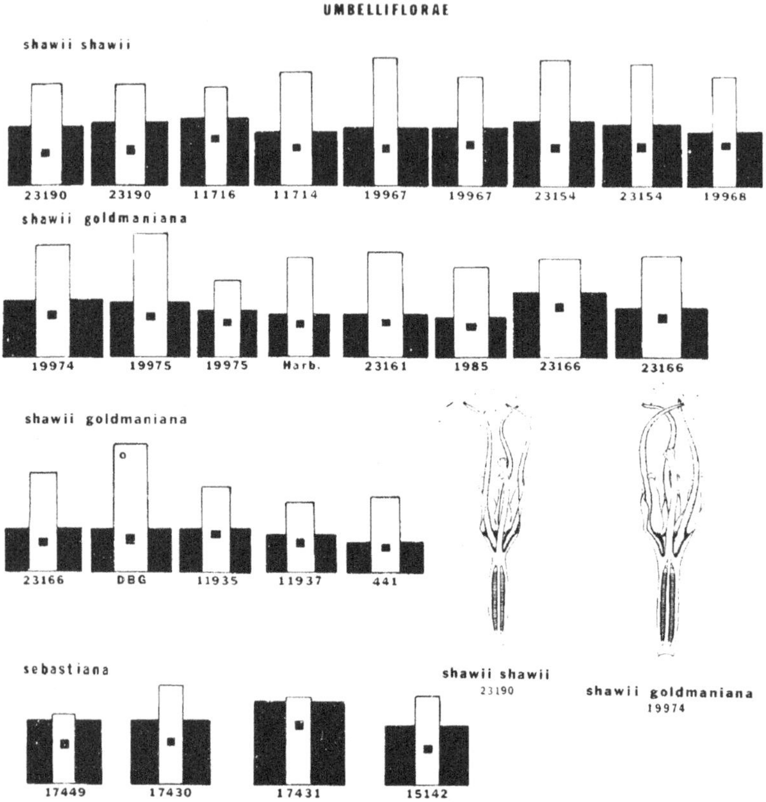

FIGURE 54. Floral ideographs of the Umbelliflorae, showing relative proportions of the tube (black) to outer tepal (white column), and level of insertion of filament (black square). Measurements are listed in Table 11.

teeth very variable in size, shape, in the mid-blade from 5 or 6 to 10–20 mm long, decreasing below, straight or variously flexed, rarely confluent or usually 1–2 cm apart, reddish to dark brown or dark gray; panicle 2–4 m tall, the peduncle stout, closely imbricated by large, purple, succulent bracts, closely investing the lateral umbellate branches; lateral branches commonly 8 to 14, horizontal to ascending, broad, stout, elliptic to oval in cross section, greenish red to purplish; flowers 75–100 mm long, sessile to short-pedicellate, bracteolate, closely clustered; buds frequently purplish or red; ovary greenish 35–50 mm long, thick-fleshy, cylindric, with grooved, scarcely constricted neck 6–15 mm long, shortly tapered at base; tube amply funnelform, 12–16 mm deep, 16–22 mm wide, thick-fleshy, knobby at filament insertions with furrows between, light yellow, smooth, shiny, sinuses sometimes at unequal levels; tepals unequal, broad and thick at base, becoming linear-lanceolate, thinner, yellow or

FIGURE 55. *Agave shawii shawii* near La Misión: (*Left*) Flowering rosette and (*right*) detail of the large, purplish, clasping bracts. Photographs by Bruce Gentry.

reddish, soon involute and wilting while stamens extend, the outer tepal longer, 25 × 5 to 38 × 8 mm, rounded to plane on outer face, the inner with fleshy keel and thin involuting margins; filaments thick, strong, 60–70 mm long when extended, inserted near mid-tube 8–12 mm above base of tube, tapered towards apex; anthers large, 20–28 mm long, usually centrically affixed; capsule variable in shape and size, 5.5 × 1.5 cm to 7 × 2.3 cm, obovoid or pyriform to oblong, slightly attenuate to rounded at base, thickly succulent when green and drying as a scurfy exocarp, with a short to long (1.5 cm) beak; seeds 7 × 5–6 mm, lunate to cuneiform, dull black to lustrous black.

Type.—*Hitchcock s.n.* 20 miles [32 km] S of San Diego on the mesa, near the ocean, at the initial monument of the Mexican boundary, July 1875, MO [4 lvs. fls., 2 sheets].

Habitat and Characteristics.—The habitat of ssp. *shawii* is a narrow one along the maritime northwest coast of the peninsula. Subspecies *shawii* is a frequent component of the coastal sagebrush community with regular associates of *Simmondsia chinensis*, *Rosa minutifolia*, *Opuntia*, *Franseria*, *Atriplex*, and other endemics like *Aesculus parryi* and *Bergerocactus emoryi*. Ten to twenty miles [16 to 32 km] inland *shawii* also contacts the higher chaparral community, where open slope and dispersed cover allow light for it to grow. Annual average rainfall ranges from

FIGURE 56. *Agave shawii* ssp. *shawii*. (*Left*) With *Idria columnaris* about 15 miles (ca. 24 km) east of Rosario. (*Right*) A mature single rosette near La Misión, about 27 miles (ca. 43 km) north of Ensenada. Photographs by Bruce Gentry.

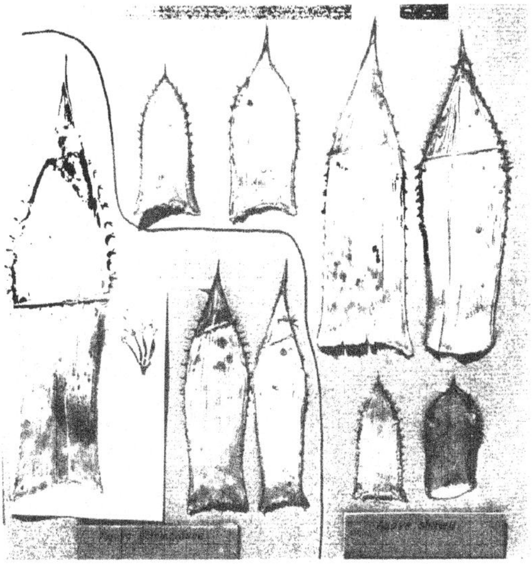

FIGURE 57. (*Upper*) Comparison of leaf series of (left) *Agave shawii goldmaniana* from the mid-peninsula, and (right) *Agave shawii shawii* from near La Misión and east of El Rosario, the two larger indicating genes of *goldmaniana*. (*Lower*) *Agave shawii goldmaniana* in massive population north of Punta Prieta in mid-peninsula.

5 to 10 inches [130–250 mm], fogs and dew are frequent, solar intensity is reduced, and temperatures are equable, with mild and infrequent frosts (Fig. 52). The bright and gleaming rosettes are especially conspicuous on the low terraces and slopes overlooking the sea (Fig. 55).

This habitat contrasts strongly with the more arid and interior habitat of ssp. *goldmaniana* (Fig. 58). These two subspecies are regarded as ecotypes of their respective habitats, slowly and presently evolving independently in their respective centers of area. The *goldmaniana* plants around Punta Prieta, for instance, are easily separable from those of *shawii* about La Misión north of Ensenada. They are not treated as species here because of my inability to separate the two along the southern perimeters of the *shawii* area, where the more robust forms of *goldmaniana* mingle with the more compact and dwarflike forms of *shawii*. This blending of forms is viewed as a natural and long-sustained potential of *shawii* to respond robustly to environmental factors in its southern area. The serious student should note, however, that this judgment is based on gross morphological considerations, not on direct genetic evidence. Series of leaves of the two subspecies are shown in Figure 57.

Exploitation of A. shawii.—*Agave shawii* fibers were probably used, and its meristem eaten, by the Indians (Barrows 1900). During the 1950's there was a brief attempt to use the coarse fibers commercially. This project was featured in a San Diego newspaper, and I later observed where plants had been bulldozed out in piles for hauling to a decorticating machine. The fibers were relatively short and of poor quality; the operation was apparently not commercially successful, and it was discontinued. Before 1950, plants were rarely seen as ornamentals in resident gardens, but since that time they are more commonly planted in the beach developments.

South of Ensenada thousands of acres with large stands of *Agave shawii* have been destroyed, along with the rest of the coastal sagebrush community, in clearing for new agricultural lands made possible by modern well drilling. Such developments together with the spread of urban growth along the beaches and environs will continue to eliminate the wild agaves. When the human density in the *A. shawii* area reaches the dimension now existing in coastal southern California, *A. shawii* may be considered an "endangered species." A flowering colony of *A. shawii* is particularly spectacular and pleasing to the human eye. This alone is sufficient to give modern man pause and to provide means of conserving this beautiful plant for the future.

Agave shawii Engelm. ssp. **goldmaniana** (Trel.) Gentry, stat. & comb. nov.

(Figures 3, 51–53, 54g, 57, 58; Tables 10, 11)

Agave goldmaniana Trel. Missouri Bot. Gard. Rep. 22:49. 1912.

Like subspecies *shawii*, but rosettes medium to large; leaves longer, generally from 40 × 10 cm to 70 × 18 cm, more acuminate, lanceolate rather than linear-ovate; teeth also variable, commonly to 10–15 mm long in upper part of blade, sometimes paired or forked at tip, "fish-tailed," 1–3 cm apart and usually joined by a heavy corneous dark margin; spine very stout 3–4 cm long, very shallowly grooved above, brown to weathered gray; inflorescence 3–5 m tall with 18 to 25 massive laterals forming a deep wide panicle, pyramidal at top rather than rounded, peduncular bracts small and more remote than in ssp. *shawii*; flowers greenish yellow to red in bud, opening yellow, 65–96 mm long; ovary green, 30–50 mm long, the neck 5–12 mm long and scarcely constricted; tube large funnelform, thick walled, bulging and furrowed from tepal sinuses; tepals subequal, the outer 20–30 mm long, 7–10 mm wide in the middle thick, linear, ascending, evenly rounded on outer face, blunt and cucullate at tip, the inner thickly keeled, ca. 2 mm shorter, low-costate within filaments inserted above middle of tube, strong elliptic to oval in cross-section, reaching 45–70 mm in length; anthers 20–27 mm long; capsule large, thick walled, short-stipitate and short beaked, 1.6 × 5.5–6 cm to 2 × 8 cm; seeds 4 × 6 to 5 × 7 mm, with small hilar notch and narrow-winged margin (*Gentry 10355*).

Type.—*Nelson & Goldman 7151*, Tinaja Yubay, Baja California, 1905, US. This locality, about 20 miles [32 km] north of Punta Prieta, is a natural water trap, as the term "tinaja" implies, in an arroyo that floods through granitic rock. I was guided there in 1949 by a native resident of the area, whose name I have forgotten. It is a watering place known only to a few, and I found no indications of dwellings, recent or prehistoric. The name has various spellings in chronicles and on maps—"Ubi," "Yubay." My guide referred to it as Tinaja Yubay. The sand-scoured pockets in the hard bedrock were nearly filled with sand, and except for brief periods after heavy

FIGURE 58. *Agave shawii* ssp. *goldmaniana* in mid-peninsula. (*Left*) In branching form; (*right*) in simple rosette form. Compare inflorescence with that of *A. avellanidens* (Fig. 36). Photographs by Bruce Gentry, April 1972.

rains, one had to dig out the sand to reach the water. Nelson and Goldman camped and watered their stock there on September 17–19, 1905, one of the few sources of sweet water in this beautiful area of the peninsula.

Distribution and Habitat.—Agave shawii goldmaniana occupies a central part of the peninsula, between latitudes 30°30′N and 28°30′N, or from Laguna Chapala to the vicinity of Mesquitál (Fig. 52). Elevations occupied range from 5 to 700 m. Except for a few tributaries that flow into the Bahía de Los Ángeles, it is limited to the Pacific slopes. Although there are gaps and irregularities, the population, by virtue of connecting strands, is nearly continuous and very large for agaves generally. It occurs on both igneous and sedimentary rocks. The most concentrated stands and best growth development observed are in the deep granitic soils south of Sierra Calamujué in the highland valley north and east of Punta Prieta, called by Goldman (1951:98) the Valley of San Andrés. Here *goldmaniana* is a co-dominant form in the vegetation along with spectacular forests of *Idria*, *Pachycereus*, *Yucca*, and *Pachycormus*. The aridity of this central highland is ameliorated by heavy nocturnal fogs moving in from the southwest from fall to spring, and which are indicated by the epiphytic lichens and *Tillandsia* that festoon the desert trees. The following notes were made among this *Agave* population on April 8, 1951, not far from the type locality, Tinaja Yubay.

Populations and Variability.—Agave shawii goldmaniana is abundant, forming extensive colonies which are dense (up to 200–300 per acre) or thinly scattered (10–20 per acre). Large individual elongate heads or stems with leaves weigh up to 200–300 lbs (ca. 90–140 kg). Plants are caulescent with stems 3–8 dm long, with old leaves deflexing and covering stems. They are usually single, but frequently form rounded clusters, the older central rosette as much as 15 dm or more tall, the whole mound 2–3 m broad. They develop with branching from leaf axils. The whole in large specimens must weigh over 1,000 lbs (450 kg). They commonly form reclinate trunks that branch repeatedly through the years from leaf axils. Old clones become rather open and cover several square meters (Fig. 3). The old dead trunks point the way back to the original plant or to where it was. The growing terminals, each tending to continue in one direction, now appear to be separate plants. Trunks in some clones are traceable back for 6–8 m, the axes being prone except for the ascending leafy terminals. The plant is thus actually tree size, but decumbent.

Leaf is variable in form, but compared to *A. shawii*, is consistently more acuminate—a distinguishing character. The leaf may be small or large, relatively narrow or broad, narrowed toward the base or scarcely so. Teeth are also highly variable and the spine only somewhat less so. The following leaf forms outline the variants, comparatively:

- Form 1. The medium or more usual: Leaf acuminate, narrow, mostly linear to slightly spatulate; teeth along the mid-blade 10–16 mm long, slender, curved or hooked, usually 1–2 cm apart; spine slender, straight, often short. Plants mostly large.
- Form 2. Leaf broad, large, spatulate, corneous edge wide, teeth heavy, 10–15 mm long; spine thick; leaf short-acuminate. Few.
- Form 3. Leaf medium in size, between Forms 1 & 2, somewhat spatulate; corneous margin and teeth heavy, 10–15 mm long; rosettes medium size. Abundant.
- Form 4. Leaf short; rosettes small to medium; spine contorted; distal end of leaf wavy; teeth long and slender 20–30 mm long. Few.
- Form 5. Teeth very broad, often forked at tip, 15–25 mm long; rosettes small, blade spatulate. Few.

Panicle is mostly 3–5 m tall with 20–30 laterals, well spread to form a long oval outline, in contrast to the subcapitate panicle of *A. shawii*. Bracts broad with buds bulging them, but smaller than those of *A. shawii*. Flowers numerous, large, yellow at anthesis, greenish in bud, and more yellow in age. Capsules variable, large, short oblong, or long obovoid being thickest just below apex. Flower form fairly constant.

Summer rains falling through the southern part of the *goldmaniana* range in 1951 following 5–7 years of drought caused abundant flowering shoots in the late summer and fall. All of the earlier September flowers failed to set fruit, while October flowers in some cases appeared to be setting fruits. Lack of fruit set in early fall was obvious that year. However, in these same populations in previous years, fruit was observed to set well during the spring months. Is lack of fruit set in fall due to: lack of pollinators, such as migrant hummingbirds; decreasing length of daylight; decreasing temperatures? Young flowering shoots in *A. goldmaniana* and *A. shawii* in October of 1951 were the most abundant ever observed.

A cryptic member of the Deserticolae, *A. avellanidens* replaces *A. shawii goldmaniana* south and east of Rancho Mesquitál. The former can scarcely be distinguished from *goldmaniana*, so similar are the green rosettes, unless *A. avellanidens* is in bloom. At that time its smaller, nearly tubeless flowers and narrow, deep panicles are evident. The small peduncular bracts and lack of axil branching in neighboring *goldmaniana* populations further complicate field identification. It is as though the *goldmaniana* population is infused with genes of single-rosette *avellanidens* that inhibit bract growth and axil branching—a speculative inference, of course.

Food for Animals.—Leaf bases and hearts are eaten extensively by pack rats (*Neotoma*), who gnaw tunnels through the leavés around the head. *Agave shawii* was frequently noted with the terminal bud undermined and dying. Ravens apparently eat the young flowers. Claw marks, like those of fox or kitfox, on the peduncle showed how a climber had secured two or three young, tender, lateral branches by breaking them from the stalk. Cattle frequently chew the tops off young flowering shoots.

Agave sebastiana

(Figures 59, 60; Table 11)

Agave sebastiana GREENE, Bull. Cal. Acad. Sci. 1:214. 1885.
Agave shawii var. *sebastiana* (GREENE) GENTRY, Allan Hancock Pac. Exped. 13:49. 1949.
?*Agave disjuncta* TREL. Missouri Bot. Gard. Rep. 22:51. 1912.

Rosettes medium to rather large, elongate, 6–12 dm tall, closely imbricate, single or cespitose, glaucous gray or green; leaves generally 25–45 × 8–24 cm, broadly linear to ovate, short-acuminate, thick rigid, plane to slightly hollowed above, rounded below, sometimes slightly narrowed towards base, light yellowish to grayish green, bud-printed, the margin usually corneous, dark brown; teeth slender, reddish brown, the larger through mid-blade 5–10 mm long, frequently down-flexed, 1–2 cm apart, or smaller and more numerous; spine stout, black to somewhat gray, 2–3 cm long, rarely shorter, variously grooved above; inflorescence 2–3 m tall, peduncle stout, with deltoid, scarious appressed bracts; panicle short, wide-spreading, rounded to nearly flat, congested with 8 to 12 large umbels of yellow flowers in upper ¼ of shaft; flowers 70–90 mm long, opening yellow, thick-fleshy; buds green; ovary 35–55 mm long, cylindric; tube broadly funnelform, 14–20 mm deep, 18–22 mm broad; tepals 16–25 × 5–7 mm ["17–33 × 11–15"],* lanceolate, markedly cucullate and glandular floccose at tip, the inner shorter and strongly keeled; filaments 50–60 mm ["70"]* long, stout, inserted in mid-tube at 8–14 mm above bottom of tube; anthers 20–21 mm ["24–30"]* long, yellow (measurements from pressed dry specimens, see Moran numbers in Exsiccatae); capsules large, ca. 30 × 60–80 mm, beaked but scarcely stipitate; seeds large, glossy black, 7 × 11 mm.

Type.—*Greene s.n.*, Cedros Island, 1 May 1885, CAS.

Agave sebastiana is closely related to *A. shawii* as shown by the particulars of habit and flower morphology. However, *sebastiana* differs significantly in the broader flatter panicle, in the smaller, more remote, scarious peduncular bracts, and in the pale green, somewhat glaucous leaves with more slender teeth. Bud printing and the black spines are more conspicuous on *sebastiana* due to the thin glaucous covering on the leaves. Reid Moran has provided an excellent series of specimens from San Benito Islands (Fig. 59 & 60) together with a series of color slides. They show that *sebastiana* forms large clones with new rosettes starting from the bases of old stems. *Moran 15142* from Natividad Island, southeast of Cedros Island, is a small form without corneous margins, the leaf only 15–20 cm long and pale green, the floral stem only 1 m tall. The flowers, however, are quite normal for *sebastiana* except for the long stamens. Moran noted it as "occasional, near middle of Natividad Island." *A. sebastiana* is a very handsome plant which should respond well in California gardens and, perhaps, elsewhere.

Agave disjuncta Trel. was based on small live plants collected by J. N. Rose, reportedly on San Benito Island. Specimens of these plants are no longer in existence. Reid Moran informs me that he has searched the island for plants of this

---

* Moran measurements of fresh flowers.

FIGURE 59. *Agave sebastiana* on Cedros Island with light-glaucous leaves. Photograph by Reid Moran.

description without success. The name has no present application.

GROUP DATYLIONES

TRELEASE, Missouri Bot. Gard. Rep. 22:45. 1912.

Acaulescent; leaves fibrous-fleshy, stiff, straight, dagger-shaped; spine strong, grooved above, scarcely decurrent; prickles heavily triangular; scape slender; panicles narrowly oblong, with greenish tubular flowers; filaments deep-seated. Baja California Sur.

Agave datylio

(Figure 61)

Agave datylio SIMON EX WEBER, Bull. Mus. Hist. Nat. Paris 8:224. 1902.

Rosettes small to medium size, 6–10 dm tall, 10–15 dm wide, suckering freely, the rhizomes frequently elongate; leaves lance-linear 50–80 cm long, 3–4 cm wide, green to yellowish green, somewhat glaucous in youth, canaliculate above, rounded below, radiately spreading, rather rigid, nearly straight; margin nearly straight; teeth mostly 3–5 mm long, deltoid, flattened, rather blunt, dark brown, usually remote or 3–6 cm apart, more closely spaced below; terminal spine large, conical to subulate, 2.5–4 cm long, dark brown to grayish, shortly decurrent, scarcely or flatly grooved above; panicles 3–5 m tall with 8 to 15 branches of small umbels in upper half of shaft; flower greenish yellow, 40–55 mm long; ovary 20–30 mm long; tube funnelform, 5–10 mm deep; tepals 15–20 mm long, erect to ascending; filaments 35–45 mm long, inserted 4–6 mm above base of tube; anthers 15–20 mm long, yellow; capsules oblong to pyriform, 1.5–2 cm in diameter, 3.5–4 cm long, short-stipitate; seeds 7 × 6 mm.

Type.—Not designated; described from La Paz, Cape District. *A. datylio* is widely scattered in the Cape region at lower elevations on granitic sandy soils (Fig. 61).

Agave datylio has no close relatives on the peninsula. It belongs with the widespread sword-leafed group, Rigidae, whose nearest geographic member is *A. aktites* Gentry of the Sonoran-Sinaloan coast. The latter, however, is amply distinct with its narrower bluish leaves, sharply cuspid teeth, larger rotund stem, and

FIGURE 60. *Agave sebastiana* on East San Benito Island in the Cedros group. An old clone with reclining trunks. Photograph by Reid Moran.

larger flowers with deep bulging flower tube. Trelease set off *A. datylio* in a monotypic group, Datyliones, but I would presently favor its inclusion in Berger's (1915) series Rigidae along with the *A. angustifolia* alliance. However, no fresh flower material of *A. datylio* is presently available, and the whole sword-leafed group has yet to be carefully revised.

A. datylio is locally called "datiliyo" or "datilillo," while its variety *vexans* at Comondú is called "mescaliyo."

Agave datylio var. **vexans** (Trel.) I. M. Johnston. Proc. Calif. Acad. Sci. ser. 4. 12:1003. 1924.

Agave vexans TREL. Missouri Bot. Gard. Rep. 22:62. 1912.

This variety differs from the typical form of the species in its smaller rosettes and leaves, ranging from 30–50 cm long, and in its more glaucous or yellowish color. Teeth are usually remote, but specimens from Arroyo Purísima have close-set, small teeth, the larger being only 2–3 mm long. Johnston separated the variety as having shorter filaments (20–30 mm), but this does not hold, as I have specimens from near Comondú with filaments 35–40 mm long, the same as for the typical form of the species. The variety occurs on sandy soils in lower elevations on both sides of the Sierra de la Giganta (Fig. 61).

Variety *vexans* appears to be a xerophytic ecotype of the species. If grown together, the varieties may show themselves to be morphologically inseparable. I had them growing for many years in Murrieta, California, but they did not flower there, perhaps because of their shady situation.

FIGURE 61. *Agave datylio*. (*Upper*) Near La Paz. (*Lower*) *A. datylio vexans* on sandy plain 23 miles (ca. 37 km) southwest of Comondú, showing the widely spaced off-sets.

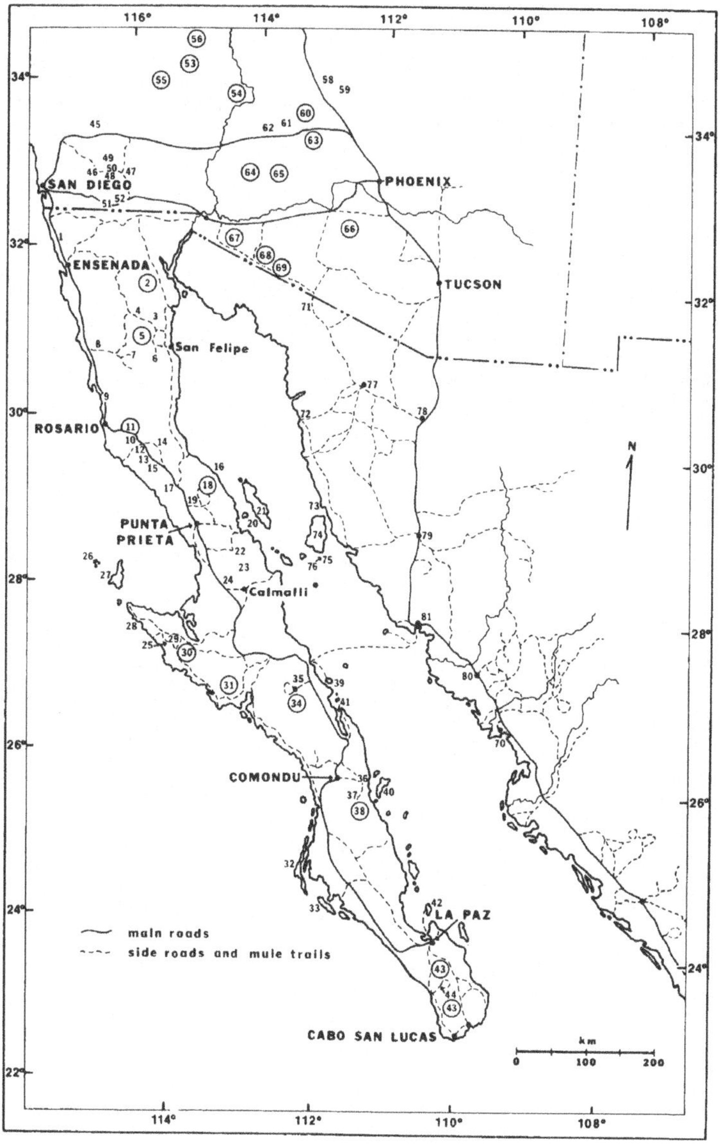

PHOENIX
TUCSON
SAN DIEGO
ENSENADA
San Felipe
ROSARIO
PUNTA PRIETA
Calmalli
COMONDU
LA PAZ
CABO SAN LUCAS
N
main roads
side roads and mule trails
km
0
100
200

APPENDIX—EXSICCATAE

Specimens that document the taxonomy, the distribution maps, and illustrations are enumerated in this index. Taxa, state names, and collectors' names are listed in alphabetical order. To facilitate these data for ready reference, the collectors' last names with their collection numbers are listed first in italics, followed by the containing herbaria in capital letters (see Abbreviations below). The localities with dates, and notes in brackets follow. Type collections and the herbaria containing holotypes are given in boldface. Curators having duplicates of these listed specimens can, therefore, easily identify their materials by using this reference. I have been unable to annotate all specimens in all herbaria, and some of my earlier identifications have been changed. Hence, this list of specimens will bring identifications up to date, provided curators use these listings. You all would do agaveology well to assign an assistant to this chore, as soon as this monograph is received.

It is not usual to include dates of collection with exsiccatae citations, but I have done this to document flowering times. The plant parts composing the collections are listed as abbreviations (see table below) in parentheses, as well as my comments on some collections. Locality citations are edited copies from collectors' labels; their appended notes are frequently given in quotes. Altogether, the data represent a record of the living agave populations and should be useful to future students.

The map of Figure 62 locates many of the little known localities listed in the Exsiccatae. Some ranches are not on maps. Other Mexican place names are frequently duplicated in a single area, such as Agua Caliente and San Felipe. Names of mountains are circled. I have found no map

---

←

FIGURE 62. Outline of the principal travels of Howard Scott Gentry in the Gulf of California region (1938–1975) and some reference localities of agave collections.

BAJA CALIFORNIA

1. La Misión	16. Bahía San Luis Gonzaga	31. Picachos de Santa Clara
2. Sierra Juárez	17. Laguna Seca Chapala	32. Isla Magdalena
3. San Matias Pass	18. Sierra San Luis	33. Isla Margarita
4. Valle de la Trinidad	19. Tinaja Yubay	34. Sierra de las Palmas
5. Sierra San Pedro Mártir	20. Bahía de Los Ángeles	35. Rancho San Sebastián
6. Rancho Agua Caliente	21. Isla Ángel de la Guarda	36. Loreto
7. Meling Ranch	22. San Borja	37. Misión San Javier
8. San Telmo	23. Paraíso	38. Sierra de la Giganta
9. Socorro	24. Rancho Mesquitál	39. Isla San Marcos
10. San Fernando	25. San Andrés	40. Isla Carmen
11. Sierra San Miguel	26. Islas San Benitos	41. Bahía Concepción
12. San Agustín	27. Isla Cedros	42. Isla Espíritu Santo
13. Catavina	28. Bahía Tortuga	43. Sierra Laguna
14. El Marmol	29. Cerro Tordillo	44. Rancho Burrera
15. Jaraguay	30. Sierra Vizcaíno	

CALIFORNIA

45. Pinyon Flats	49. Borrego Springs	53. Providence Mountains
46. San Felipe Ranch	50. Carriso Creek	54. Whipple Mountains
47. San Felipe Creek	51. Jacumba	55. Dad Granite Mountains
48. Yaqui Wells	52. Mountain Springs	56. Ivanpah Mountains

ARIZONA

58. Burro Creek	62. Indian Springs	66. Sand Tank Mountains
59. Hillside	63. Harquahala Mountains	67. Tinajas Altas
60. Harcuvar Mountains	64. Kofa (S. H.) Mountains	68. Cabeza Prieta Mountains
61. Cunningham Pass	65. Little Horn Mountains	69. Tule Mountains

SONORA

70. Yavaros	74. Isla Tiburón	78. Santa Ana
71. Quítovac	75. Turner's Island	79. Hermosillo
72. Puerto Libertad	76. Seal Island	80. Ciudad Obregon
73. Punta Tepopa	77. Caborca	81. Guaymas

that locates Sierra de las Palmas (34), yet it is a prominent extension of the Sierra de la Giganta with a local montane vegetation of oak-*Nolina* grassland. Notable additions to the peninsular flora have been found there, e.g., *Agave gigantensis* and *Sedum alamosanum* from the mainland. The main road through the peninsula represents the old unpaved auto trail. However, on a scale of this ratio, it was necessary to forego close accuracy in detail.

ABBREVIATIONS FOR HERBARIA

A = Arnold Arboretum, Harvard University, Cambridge, Massachusetts.
ASU = Arizona State University, Tempe, Arizona.
ARIZ = University of Arizona, Tucson, Arizona.
BH = Bailey Hortorium, Cornell University, Ithaca, New York.
CAS = California Academy of Sciences, Golden Gate Park, San Francisco, California.
DES = Desert Botanical Garden, Phoenix, Arizona.
DS = Dudley Herbarium, now in CAS.
GH = Gray Herbarium, Harvard University, Cambridge, Massachusetts.
MEXU = Instituto de Biología, Universidad Nacional Autónoma de México.
MICH = University of Michigan, Ann Arbor, Michigan.
MO = Missouri Botanical Garden, St. Louis, Missouri.
NA = National Arboretum Herbarium, Washington, D.C.
POM = Pomona College, Claremont, California.
SBBG = Santa Barbara Botanic Garden, Santa Barbara, California.
SD = San Diego Museum of Natural History, San Diego, California.
UC = University of California, Berkeley, California.
US = National Herbarium, Natural History Museum, Washington, D.C.

Gentry Herbarium specimens are presently on deposit in the Desert Botanical Garden Herbarium and are included in that listing, DES.

PARENTHETICAL ABBREVIATIONS

br = bract
cap = capsule
f = flower
l = leaf
photo = photograph

Agave aurea

BAJA CALIFORNIA

Barclay & Arguelles 1988, DES, MEXU, US. 5 miles [8 km] N of Todos Santos along road to La Paz, 20 Apr. 1966 [l, f].

***Brandegee s.n.*, UC,** DS. Type. Purísima, 13 Feb. 1889 [margin of large l, small l, 5 f, pieces of 2 cap. Has typically close-set, regular teeth on regular teats forming undulate margin, as growing on lava mesas about Comondú and E of Purísima].

Brandegee s.n., MO. Cape Region mountains, 20 Sept. 1899 [l, f, mixed with f of *A. sobria*; type of *A. brandegeei* Trel.].

Carter 5132, UC. Mesa de San Gerónimo, N from Rancho Viejo [on road from Loreto to San Javier], 8 May 1966; alt. ca. 1,110 m.

Carter 5484, UC. Arroyo de Puerta Vieja on road from Loretto to Comondú, Sierra de la Giganta, 4 July 1970; alt. ca. 1,400–1,500 ft [ca. 430–460 m].

Carter 5779, UC. Mesa de Humi, a mesa on crest of Sierra de la Giganta opposite N end of Isla San José, 20 Mar. 1973; alt. ca. 750 m.

Gentry 11299, DES. Polymorphic population on mesa NW of Comondú [photo].

Gentry 12341, DES, MEXU, US. 5 miles [8 km] N of Todos Santos, Cape District, 22 Nov. 1952 [l, f].

Gentry 12375, DES, MEXU, US. 3 miles [5 km] N of Comondú, 30 Nov. 1952 [f].

Gentry 10321, DES, MEXU, US. Comondú, 2 Apr. 1951 [l, f, cap].

Gentry 11253, DES, MEXU, US. Ca. 10 miles [ca. 16 km] W of San José del Cabo, 5 Oct. 1951 [l, photo].

Gentry 11198, DES, US. Ca. 12–15 miles [ca. 20–25 km] E of La Paz, Cape District, 29 Sept. 1951 [l, photo].

Gentry 4272, ARIZ, DES, DS, US. Cerro de la Giganta, 1 Mar. 1939 [f]. Volcanic slopes & crags up to 4,000 ft [ca. 1,200 m].

Gentry 12383, DES, MEXU, US. Ca. 3 miles [ca. 5 km] N of Misión San Javier, Sierra Giganta, 2 Dec. 1952.

Gentry & Cech 11255, DES, MEXU, US. Las Cuevas, Cape District, 7 Oct. 1951 [l, photo].

Gentry & Cech 11283, DES, MEXU, US. 35

miles [ca. 56 km] NW of La Paz, 11 Oct. 1951 [l, photo].

Gentry & Cech 11295, DES, MEXU, US. 3 miles [ca. 5 km] NW of Comondú, 18 Oct. 1951 [l, photo].

Gentry & Cech 11301, DES, MEXU, US. 10 miles [ca. 16 km] W of Canipolé along road to Purísima [l, photo].

Gentry & Gentry 23182, DES, MEXU, US. 3 miles [ca. 5 km] N of Comondú on lava fields, 10 Apr. 1973 [l, f, photo].

Harbison s.n., SD. Mesa 2 miles [ca. 3 km] N of Comondú; elev. 400 m, 7 Oct. 1967.

Harbison s.n., SD. 6 miles [ca. 10 km] N of Todos Santos.

Moran 7144, CAS, DS, MEXU, SD. Cape District, 4 km N of La Huerta, 400 m, 25 Jan. 1959 [l, f].

Purpus s.n., MO, UC. San José del Cabo, Jan.–Mar. 1901 [l, f].

Wiggins 14470, CAS, DS, MEXU. 100 miles [ca. 160 km] W of Los Planes, alt. 490 m, south-facing slopes, 21 Dec. 1958 [l, f].

Wiggins 14531, CAS, DS, MEXU, UC. 6.4 miles [10.3 km] N of Todos Santos along road to La Paz, 25 Dec. 1958 [l, f].

Agave avellanidens

BAJA CALIFORNIA

***Brandegee s.n.*, UC.** Type. Paraíso, 1 May 1889 [upper ½–⅔ of leaf, flowers, and old caps. All similar to what I have seen and collected between Calmallí and Mesquitál Rancho].

Gentry & Fox 11944, DES, MEXU, US. Ca. 9 miles [ca. 30 km] E of Punta Prieta, 19 May 1952.

Gentry & Fox 11933, 11932, DES, MEXU, US. 6 miles [ca. 10 km] W of Calmallí, 17 May 1952.

Gentry & Fox 11929, DES, MEXU, US. 6 miles [ca. 10 km] W of Calmallí, 17 May 1952 [l, f, photo].

Gentry & Gentry 23184, DES, MEXU, US. 5½ miles [ca. 9 km] W of Calmallí, 12 Apr. 1973 [l, f, photo].

Gentry & Gentry 23186, DES, MEXU, US. 15–17 miles [ca. 24–27 km] SE of Mesquitál, 12 Apr. 1973 [l, f].

Gentry & Gentry 23187, DES, MEXU, US. 7 miles [ca. 11 km] SE of Mesquitál along road to El Arco, 12 Apr. 1973 [l, f].

Hammerly 69, CAS. 29 miles [ca. 47 km] N of Mesquitál, 27 Sept. 1941 [doubtfully referred here].

Wiggins 5726, DS. Between Calmallí and Mesquitál Rancho on mesas, 31 May 1931 [l,f].

Agave capensis

BAJA CALIFORNIA

Barclay & Arguelles 1987, MEXU, US. Vicinity of Cabo San Lucas, 19 Apr. 1966.

Brandegee s.n., UC. Cabo San Lucas, 18 Mar. 1892 [l, cap].

Gentry 10080, DES, MEXU, US. Huntington Botanical Gardens, San Marino, California, 9–15 Jan. 1951 [l, cap, photo].

Gentry 19676, DES, MEXU, US. Huntington Botanical Gardens, 17 Apr. 1962 [l, f].

Gentry & Fox 11247, *11250*, DES, MEXU, US. Type. Cabo San Lucas & vicinity, 5 Oct. 1951 [l, f, photo].

Gentry & Fox 11823, DES, MEXU, US. 3 miles [ca. 5 km] N of Todos Santos, Cape Dist., 4 May 1952.

Agave cerulata cerulata

BAJA CALIFORNIA

Brandegee s.n., DS. Cardón Grande (between San Ignacio & Calmallí), 22 Apr. 1889 [l, f, cap].

Carter & Kellogg 2953, UC. Isolated red hill in sandy plain, 18.2 km W of Misión Santa Gertrudis, 18 Dec. 1950.

Ferris 8576A, DS. 1 mile [ca. 1.6 km] S of Laguna Seca Chapala, 6 Mar. 1934 [l, cap].

Gentry 10346, DES, MEXU, US. Calmallí, between mine and houses of town, 6 Apr. 1951, type locality.

Gentry 10359, DES, MEXU, US. Ca. 7 miles [ca. 11 km] S of Tinaja Yubay and 15 miles [ca. 24 km] NE of Punta Prieta; 8 Apr. 1951 [l, cap].

Gentry 10369, MEXU, US. Rancho Jaraguay, 9 Apr. 1961 [l, early f, eaten by cattle].

Gentry 11188, DES, MEXU, US. 4 miles [ca. 6 km] NW of Laguna Seca Chapala, 21 Sept. 1951 [l, cap].

Gentry 19973, DES, MEXU, US. 10 miles [ca. 16 km] S of Laguna Seca Chapala, 29 Apr. 1963 [l, f, photo].

Gentry & Cech 11322, DES, MEXU, US. 4 miles [ca. 6 km] NW of Laguna Seca Chapala, 24 Oct. 1951 [l, photo].

Gentry & Fox 11919, 11921, 11924, DES, MEXU, US. 6 miles [ca. 10 km] W of Calmallí, 17 May 1952.

Gentry & Fox 11953, DES, MEXU, US. 21 miles [ca. 34 km] E of Punta Prieta on road to Bahía de Los Ángeles, 20 May 1952 [l, photo]

Gentry & Fox 11961, 11962, DES, MEXU, US. 4 miles [ca. 6 km] NW of Laguna Seca Chapala, May 1952 [l, photo].

Gentry & Gentry 23159, DES, MEXU, US. 20 miles [ca. 32 km] SE of San Agustín, elev. 2,100 ft [ca. 640 m], 5 Apr. 1973 [l, early f].

Gentry & Gentry 23185, DES, MEXU, US. 6½ miles [ca. 10 km] W of Calmallí, 12 Apr. 1973.

Gentry & McGill 23298, DES, MEXU, US. 10 miles [ca. 16 km] S of San Luis Gonzaga Bay along road to Laguna Chapala, 17 June 1973 [l, f].

Gentry & McGill 23302, DES, MEXU, US. W. of Sierra Calamajué, ca. 30 miles [ca. 48 km] N of Punta Prieta, 17 June 1973 [l, f, photo].

Gentry & McGill 23306, DES, MEXU, US. 16–20 miles [ca. 26–32 km] NE of Punta Prieta along road to Bahía de Los Ángeles, 17 June 1973 [l, f].

Gentry & McGill 23307, DES, MEXU, US. Ca. 40 miles [ca. 65 km] NE of Punta Prieta along road to Bahía de Los Ángeles, 17 June 1973 [l, f].

Gentry & McGill 23314, DES, MEXU, US. 6–8 miles [ca. 10–13 km] S of road fork from Los Angeles Bay along road to San Borja, 18 June 1973 [l, f].

Harbison s.n., SD. 12 miles [ca. 19 km] E of Calmallí, 8 Apr. 1947 [l, f].

Harbison s.n., SD. 20 miles [ca. 32 km] S of Punta Prieta, 9 Apr. 1947 [l, inflo].

Harbison s.n., SD. Agua Amarga, ca. 15 miles [ca. 24 km] W of Los Angeles Bay, 15 Apr. 1947 [l, f].

Johnston 3487, 3489, CAS, GH, SD, UC, US. Los Angeles Bay, 6 May 1921, "in small groups on rocky mountainside," [yellow leaves with brown-ringed teeth, l, f, cap].

Johnston 3405 a–g, CAS, US. Angel de la Guarda Island, Palm Canyon, 3 May 1921 [a series of yellow leaves, mostly with reduced teeth, brown-ringed, and some toothless], "gregarious on hillside."

Moran 4106, BH, DS. Motherless Island, Los Angeles Bay, 10 May 1952 [l, cap].

Moran 2007, DS, UC. 16 miles [ca. 26 km] N of Punta Prieta, elev. ca. 1,700 ft [ca. 520 m], 22 Apr. 1946 [l, f].

Moran 8167, DS, SD, UC. 3 miles [ca. 5 km] E of El Arco, elev. ca. 250 m, 5 Apr. 1960 [l, f].

Moran 8185, SD. Arroyo Estatón, Isla Ángel de la Guarda, 15 Apr. 1960; elev. ca. 25 m [f, cap, doubtfully assigned here].

***Nelson & Goldman 7180*, US.** Type. Calmallí, elev. 800 ft [ca. 240 m], 29 Sept. 1905. [Leaves and flowers characteristic of what I have seen through middle Baja California; no problem here.]

Raven et al. 12631, UC. 1.6 km S of Rancho Santo Ignacito, elev. 560 m, 21 Apr. 1958.

Stover & Harbison s.n., SD. 35 miles [ca. 56 km] N of Punta Prieta, 5 May 1939 [l, f].

Thomas 7972, SD, US. 1 mile [ca. 1.6 km] NW of Pozo Alemán, elev. ca. 800 ft [ca. 240 m], 26 May 1959 [l, bud].

Wiggins 5721, DS. 15 miles [ca. 24 km] NW of San Ignacio, 30 May 1931 [l, f].

Wiggins 5724, DS, UC, US. Calmallí, 31 May 1931 [l, f].

Wiggins 5734, DS. 5 miles [ca. 8 km] N of Punta Prieta, 1 June 1931 [l, f].

Wiggins & Wiggins 14882, CAS, DS. S end of Isla Ventana, Bahía de los Ángeles, near beach, 18 May 1959 [l, cap].

Agave cerulata dentiens

BAJA CALIFORNIA

Bostic s.n., SD. San Esteban Island, sandy arroyo near SE corner, 21 June 1965.

Johnston 3194, CAS, UC. San Esteban Island, 20 Apr. 1921, "common in small colonies on hillsides," [has brown-ringed teeth like *cerulata*, l, cap].

Moran 4079, SD. San Esteban Island, 6 May 1952.

Moran 21748, SD. San Esteban Island, arroyo near E side, Apr. 1975, "large colonies on hillsides and in arroyo" [in bud].

***Rose 16819*, US.** Type. San Esteban Island, 12 Apr. 1911 [l, cap].

Agave cerulata nelsonii

BAJA CALIFORNIA

Gentry et al. 10370, 11155, 11162, 11164, 11165, 11665, 11666, DES, MEXU, MICH, US. 2–3 miles [ca. 3–5 km] N of San Fernando, Sierra San Miguel, type locality, 9 Apr. 1951, 10 Sept. 1951, 10 Apr. 1952 [l, f, cap].

Gentry et al. 10376, DES, MEXU, US. 18 miles [ca. 30 km] E of Rosario, 10 Apr. 1951.

Gentry et al. 11178, DES, MEXU, US. 28 miles [ca. 45 km] E of Rosario on Sierra San Miguel, 13 Sept. 1951.

Gentry et al. 11179, DES, MEXU, US. 17 miles [ca. 27 km] E of Rosario, 14 Sept. 1951 [l, f].

Gentry et al. 11185, DES, MEXU, US. Ca. 4 miles [ca. 6 km] SE of San Agustín, elev. ca. 2,000 ft [ca. 3,200 m], 21 Sept. 1951 [l, f].

Gentry & McGill 23311, DES, MEXU, US. 4 miles [ca. 6 km] S of Los Angeles Bay road fork on road to San Borja, elev. 900 ft [ca. 275 m], 18 June 1973 [l, f].

Gentry & McGill 23315, DES, MEXU, US. 10 miles [ca. 16 km] N of San Borja, elev. ca. 1,700 ft [ca. 500 m], 18 June 1973 [l, f].

Gentry & McGill 23322, DES, MEXU, US. Ca. 5 miles [ca. 8 km] N of San Fernando, Sierra San Miguel, elev. ca. 1,750 ft [ca. 530 m], 22 June 1973 [l, f].

Gentry & McGill 23324, DES, MEXU, US. 11 miles [ca. 18 km] E of Rosario, Rancho Porvenir, elev. 450 ft [ca. 140 m], 24 June 1973 [l, f].

Harbison s.n., SD. 1 mile [ca. 1.6 km] W of Rancho Arenoso; elev. ca. 500 m (?).

Moran 22643, SD. S of Rancho San Miguel [Sierra San Miguel Range], elev. ca. 900 m, 9 Aug. 1975 [f].

Moran & Reveal 22064, SD. Ridge 3 miles [ca. 5 km] SW of San Isidro, 30°44′N, 115°34′ W, elev. ca. 1,120 m, 20 July 1975, "occasional in chaparral" [l, f].

Nelson & Goldman 7111, **US**. Type. San Fernando [Sierra San Miguel], alt. 1,400 ft [ca. 425 m], 4 Sept. 1905 [l, old f].

Agave cerulata subcerulata

Baja California

Barclay & Arguelles 1991, DES, MEXU, US. Ca. 5 miles [ca. 8 km] W of San Ignacio, 29 Apr. 1966 [l, f].

Gentry 10330, DES, MEXU, **US**. Type. San Ignacio, 3 Apr. 1951 [l, f, cap, inflo].

Gentry 11892, DES, MEXU, US. Arroyo de la Tenería, Isla San Marcos, 13 May 1952 [l, f, cap, photo].

Gentry & Fox 11926, DES, MEXU, US. 6 miles [ca. 10 km] W of Calmallí, 17 May 1952.

Gentry & Gentry 23170, DES, MEXU, US. 10 miles [ca. 16 km] W of San Ignacio, 8 Apr. 1973 [l, f, photo].

Gentry & Gentry 23175, DES, MEXU, US. 24 miles [ca. 40 km] E of San Ignacio along road to Santa Rosalia, 9 Apr. 1973 [l, f].

Harbison No. P, DES. Cuesta de las Vírgenes, 1972.

Johnson 3649, 3650, CAS, GH, SD, UC, US. San Marcos Island, on gypsum, 12 May 1921 [l, f, cap].

Pinkava & McGill P12287, ASU, DES. Ca. 70 miles [ca. 110 km] W of Santa Rosalia, Route 1. 30 May 1974. Butte base [l, f].

Agave datylio

Baja California

Brandegee s. n. CAS. La Paz, Cape District, 4 Nov. 1891 (l, f).

Brandegee 581, UC. San Pedro, 29 Oct. 1891 [l, f, cap].

Brandegee s. n. UC. Paseo de los Dolores to Lake Ramon, 4 Apr. 1889 (l, cap).

Gentry 11200, DES, MEXU, US. About 4 miles [ca. 6 km] E of La Paz, Cape District, 29 Sep. 1951 [l, f, photo].

Moran 3553, SD. Ensenada de los Muertos, Cape District, 1 Apr. 1952 [l, cap].

Peters 124, UC. Los Planes, Cape District, arroyo bottom, elev. 300 ft [ca. 90 m], 27 Mar. 1948, "mescal."

Rose 1302, US. La Paz, 14 June 1897.

Wiggins 11501, DS. 9 miles [ca. 15 km] W of La Paz.

Wiggins 15475, CAS, DS, MEXU. Rancho de Obispo [Magdalena Plain], alt. ca. 150 m, 15 Nov. 1959 [l with long blade of *A. datylio* and remote teeth of *vexans*].

Agave datylio var. **vexans**

Baja California

Brandegee s. n. UC. Purísima?, 1899 [infl]

Gentry 10302, DES. Rancho Panales, Arroyo Purísima, 31 Mar. 1951 (l, f, cap). Rocky sedimentary slope; cardon-pitaya-Bursera-Larrea-Opuntia-etc.

Gentry 4322, DES. Comondú, 10 Mar. 1939 (f). Hill slope, common over coastal plain and foothills.

Gentry & Cech 11292, DES. 23 miles [ca. 37 km] SW of Comondú, 16 Oct. 1951 (l, photo) Sandy valley bottomland.

Hastings & Turner 64–377, ARIZ, CAS. 6 miles [ca. 10 km] by road W of San Luis Gonzaga (NW of La Paz), elev. 400 ft [ca. 120 m], 2[illegible] Oct. 1964, flowers yellow with green tinge [l, f]

M. E. Jones 23751, MO. La Paz, 15 Nov. 192[illegible] [f].

Rose 1302, US. La Paz, 14 June 1897.

Rose 16540, US. La Paz, 29 Mar. 1911.

Wiggins 15205. CAS. Mesa 2 miles [ca. 3 km N of Arroyo San Gregorio, 27 Oct. 1959 [l, f]

Agave deserti deserti

BAJA CALIFORNIA

Gentry s.n., DES. Granitic sandy highland on Mex. Rt. 2, S of Jacumba, Apr. 1963 [photo].

Gentry & Arguelles 22990, DES, MEXU, US. Valle de Trinidad, elev. ca. 2,600 ft [ca. 800 m], 3 May 1972 [l, f].

Gentry & McGill 23285, DES, MEXU, US. 17–18 miles [ca. 27–29 km] W of San Felipe along road to Valle Trinidad, elev. ca. 1,950 ft [ca. 600 m], 13 June 1973 [l, f].

Gentry & McGill 23286, ASU, DES, MEXU, US. 2.6 miles [ca. 4 km] SE of Rancho Agua Caliente on E bajada of Sierra San Pedro Mártir, elev. ca. 1,450 ft [ca. 440 m], 13 June 1973 [l, f].

Hastings & Turner 66-6. ARIZ. 12.6 miles [ca. 20 km] W of turnoff toward San Matias Pass, elev. 1,300 ft [ca. 400 m], 4 Oct. 1966 [l, cap].

Hutchison 710, UC. Between Alaska and Mexicali, km 140, E slopes of the sierra, 31 Dec. 1952 [flowered UC Botanical Garden, Aug. 1963; karyotype n = 59, Cave].

Pinkava & McGill 8648, ASU, DES. Along Rte. 2, 38.5 miles [ca. 62 km] W of junction with main route to Mexicali, 7 June 1971 [l, f, br].

Wiggins & Wiggins 16044, DS. Granitic sandy bajada 16 miles [ca. 26 km] W of San Felipe Hwy. along road to San Matias Pass, 2 Apr. 1960 [l, f].

CALIFORNIA

Abrams 3976, DS. Between San Felipe & Carisso Creek, San Diego Co., 4 July 1903 [l, f].

Ball & Everett 22777, UC. Ca. 3 miles [ca. 5 km] NE of Banner Trading Post, State Hwy. 78, 6 Nov. 1957 [l, cap].

Ball & Everett 22671, UC. 6.6 miles [ca. 11 km] SW of Palm Village Hwy. junction 111 along Hwy. 74, elev. 1,650 ft [ca. 500 m], 20 Aug. 1957 [toothless l and cap].

Barr 67-211, 67-210, ARIZ. 40 miles [ca. 65 km] W of El Centro near Ocotillo, 13 May 1967; elev. 1,700 ft [ca. 520 m] [l, f].

Cleveland 7218, SD. Jacumba, 25 June 1885 [f].

Crovello 295, UC. 2.6 miles [ca. 4 km] E of Vallecito Stage Station, San Diego Co., elev. 2,300 ft [ca. 700 m], 15 Dec. 1963 [f].

Eastwood 18639, CAS. Road to Mountain Springs, San Diego Co., 25 Apr. 1932.

Ferris & Rossbach 9685, DS. Borrego Springs, P. O. road and State Hwy. 78, San Diego Co., May 1938 [l, f].

Ferris 7060, DS. Between Jacumba & Mt. Springs, San Diego Co., 18 Apr. 1928 [l, cap].

Flemming 753, SD. Foot of Banner Grade, Sentenac Canyon, [San Diego Co.] Apr. 1926.

Gander 1325, SD. Carriso Creek, [San Diego Co.] 3 Jan. 1936 [l, f].

Gander 4810, SD. Bull Willow Canyon, San Diego Co., 15 Dec. 1937 [l, f].

Gentry 10034, 10034a, DES, MEXU, MICH, US. 3–4 miles [ca. 5–6 km] S of Palm Desert, Santa Rosa Mt., 20 Dec. 1950.

Gentry 10041, 10044, DES, MEXU, MICH, US. Yaqui Wells near San Felipe Creek, San Diego Co., 29 Dec. 1950 [l, cap].

Gentry 10051, DES, MEXU, US. Pinyon Flats, Hwy. 74, Riverside Co., elev. ca. 4,000 ft [ca. 1,200 m], 5 Jan. 1951 [l, f, cap].

Gentry 17759, DES, MEXU, US. Pinyon Flats, [Riverside Co.], 7 July 1959.

Gentry 19741, DES, MEXU, US. San Felipe Ranch near Rt. 78 and road to Warner's Ranch, San Diego Co., 25 May 1962 [l, f].

Gentry et al. 11650, 11652, DES, MEXU, MICH, US. Pinyon Flats, [Riverside Co.], 3 Apr. 1952.

Gentry & McGill 23326, DES, MEXU, US. Pinyon Flats, [Riverside Co.], 26 June 1973.

Hall 2117, DS, UC, US. E base of San Jacinto Mts., Colorado Desert, June 1901 [l, f].

Hitchcock & Palmer in 1875, ***Emory*** in 1846, **MO**. "The types." E of San Felipe Ranch, San Diego Co. [Probably along San Felipe Creek, above or below The Narrows.]

Howell 3243, CAS. San Felipe Wash, halfway from Borrego Valley to Yaqui Wells, San Diego Co., 26 Nov. 1927 [l, f, cap].

Huey s.n., SD. Borrego Narrows, [San Diego Co.], Apr. 1931 [f].

Lester s.n., SD. San Isidro Mt., N of Borrego, May 1926.

Mearns 2972, US. E base of Coast Range, edge of Colorado Desert, 7 May 1894.

Mearns 3109, 3025, 3147, DS, US. Mountain Springs, San Diego Co., May 1894 [l, inflo].

Palmer 462, 88, GH. Type collection in part. San Felipe Canyon, San Diego Co., 1875 [inflo].

Vasey 626, US. Mountain Springs grade, [San Diego Co.], June 1880 [l, f].

Woglum 156, SD. Mountain Springs grade [San Diego Co.], 7 May 1936.

Wolf 9455, UC. Palms to Pines Hwy. 74, 2.7 miles [ca. 4 km] below Dos Palmos Spring, Riverside Co., 29 Sept. 1939 [l, f].

Woodcock s.n., SD. Foot of Banner Grade, Sentenac Canyon [San Diego Co.], 10 Apr. 1929 [f].

Yates 5459, UC. 5 miles [ca. 8 km] W of San Felipe Canyon, San Diego Co., elev. 1,500 ft [ca. 460 m], 8 Apr. 1936 [l, f, cap].

Agave deserti pringlei

BAJA CALIFORNIA

Broder 547, 473, DS. 3 miles [ca. 5 km] WNW of Santa Catarina, 25 May 1961, elev. ca. 4,000 ft [ca. 1,200 m] [l, f. In Sierra Juárez].

Gentry 10287, DES, MEXU, US. Northwest end of Sierra San Pedro Mártir, 23 Mar. 1951.

Gentry 16723, DES, MEXU, US. Near San Matias Pass, 22 June 1957 [l, f, photo].

Gentry 19959, DES, MEXU, US. San Matias Pass, 23 Apr. 1963 [l, f, photo].

Harbison s.n., SD. Near El Progreso, Sierra Juárez, elev. ca. 1,500 m; 32°18′ N, 115°54′ W, 1 Aug. 1965 [l, f].

Hastings & Turner 66–17, ARIZ. San Matias Pass, 19.1 miles [ca. 31 km] E of Valle Trinidád, 5 Oct. 1966, elev. 1,950 ft [ca. 600 m] [l, cap].

Moran 9838, SD, UC. 5 miles [ca. 8 km] SE of Las Filipenas, Sierra Juárez, elev. ca. 1,620 m, 30 June 1962 [l, f].

Moran 9849, SD. Just E of San Matias Pass, elev. ca. 1,020 m, 30 June 1962 [l, f, caps]. "Rosettes clustered, to 1.2 m, of ca. 50 leaves; leaves green, channeled, curved, to 7 dm long, floral stem 3–5 m tall . . . ."

Moran 15256, SD. Rancho San Pedro Mártir, Sierra San Pedro Mártir, elev. ca. 1,700 m, 5 July 1968 [l, f]. "Rosettes 1–1½ m, of ca. 50 green or slightly glaucous leaves 4–8 dm long. Occasional in chaparral."

Moran 18639, SD. 2 miles [ca. 3 km] NE of Alamito, Sierra Juárez, elev. ca. 1,150 m, 3 Oct. 1971 [l, f].

Moran 21983, SD. Sierra San Pedro Mártir, 30°57′ N, 115°36′ W, elev. ca. 1,600 m, "on metamorphic rock in small arroyo, mile NW of oak pasture" [l, cap].

Moulis & McGill 555, ASU, DES. Sierra San Pedro Mártir, ca. 18.5 road miles [ca. 30 km] E of Meling Ranch, 21 Aug. 1972 [l, f, cap].

Orcutt s.n., UC. Hanson's Ranch, 29 July 1883.

Orcutt s.n., DS. Lower California, 1892, ex. herb. Pringle. [Sierra Juárez.]

Orcutt s.n. **K**, MEXU. Type. Sierras Centrales, elev. 6,000 ft [ca. 1,830 m], 7 Oct. 1882 [l, cap].

Agave deserti simplex

ARIZONA

Engard s.n., DES. Near Indian Springs, Harcuvar Mt., Yuma Co., June 1974 [f only].

Gentry 23404, DES, SD, **US**. Type. N slope of Harquahala Mt., 12 miles [ca. 19 km] W of Águila, Yuma Co., 12 June 1974 [l of 5 & f of 4 pls.].

Gentry 20590, DES, MEXU, US. Near a sheep tank in Cabeza Prieta Mts. [Yuma Co.], 3 May 1964.

Gentry & Engard 23562, ASU, DES, US. S end of Silver Bell Mts., Pima Co., 17 June 1975. Limestone, elev. 3,000–3,500 ft [ca. 900–1,070 m], [l, f].

Gentry & Ogden 9947, MEXU, US. N. slope Harquahala Mt., elev. 3,500–5,000 ft [ca. 1,000–1,500 m], 10 Nov. 1950 [l, f].

Gentry & Ogden Photo, DES. Cunningham Pass, Harcuvar Mts., Yuma Co., 14 Nov. 1950.

Gentry & Weber 23410, DES. Sand Tank Mt., Yuma Co., 27 July 1974; rocky volcanic slope, elev. 2,800–3,000 ft [ca. 850–900 m], [l, cap, 2 live].

Goldman 2310, US. Tinaja Altas, Yuma Co., 20 Nov. 1913.

Harbison 4312, ARIZ, SD. Canyon leading to Del Oso Pass, Kofa Mts., Yuma Co., 8 June 1943 [The largest leaf seen of this species, with large teeth] [l, f].

Harrison et al. 7303, ARIZ. Table Top Mt., 16 Aug. 1930 [l, cap]. [Doubtfully referred here.]

Mearns 305, US. Tule Mts., Mexican boundary line, 11 Feb. 1894 [l looks like *A. zebra*].

Peebles & Smith 14415, ARIZ, GH, US. Sierra Estrella, Maricopa Co., elev. 2,500 ft [ca. 760 m], 12 July 1939 [l, cap].

Peebles & Smith 13873, 13878, 13881, ARIZ, GH, US. Cunningham Pass, Harcuvar Mts., Yuma Co., 17 May 1938 [l, f, cap].

Van Devender s.n., ARIZ. Along Hwy. I 8 in Telegraph Pass, Gila Mts., N slope, elev. ca. 800 ft [ca. 250 m], 31 Dec. 1972 [l, cap].

Weber Photo, DES. Summit of Little Horn Mts., May 1970.

Weber & McGill 2549, ASU, DES. ½ miles [ca. 0.8 km] S of Sheep Tank Mine, Yuma Co., Ariz., 1 June 1970 [l, f].

West s.n., ARIZ. 18-Mile Drive, Organ Pipe

Cactus National Monument, Pima Co., 27 May 1962 [l, f].

Wiggins 8643, ARIZ, DS, GH, UC, US. Rocky canyon near small tanks ca. 10 miles [ca. 16 km] S of Hwy. 80 on E foot of Mohawk Mts., Yuma Co., 1 Mar. 1937 [l, cap—Atypical elongate leaf; flowers should be collected].

CALIFORNIA

Blakley 3231A, DES. Above Mitchell's Caverns, Providence Mts. [flowered Santa Barbara Botanic Garden, 22 June 1972].

Brandegee s.n., UC. Providence Mts., 31 May 1902, "plants solitary."

Ferris & Bacigalupi 8161, DS. Canyon above Bonanza King Mine, Providence Mts., San Bernardino Co., 27 Apr. 1932.

Jaeger s.n., SD. 20 miles [ca. 32 km] NE of Amboy, E end of Granite Mts., [San Bernardino Co.], 26 Mar. 1929.

Munz et al. 4302, US. Vicinity of Bonanza King Mine, E slope of Providence Mts., Mohave Desert, alt. 3,500–5,000 ft [ca. 1,070–1,500 m], 21–24 May 1920 [cap].

Shaw s.n., DES. Ivanpah Mt. (Riley's claim), San Bernardino Co., Spring, 1974 [l, cap].

Van Devender et al. 74-29, ARIZ. S slopes of Whipple Mts., San Bernardino Co., elev. 1,600 ft [ca. 500 m], 17 Feb. 1974 [l, f].

Wolf 10183, DS, US. N side of Old Dad-Granite Mt. Range, Snake Spring, Mohave Desert, elev. 4,500 ft [ca. 1,400 m], 30 Apr. 1941.

Wolf 3165, CAS, DS, UC. From Copper Basin on road to Parker Bridge, Whipple Mts., E Colorado Desert, San Bernardino Co., elev. 1,400 ft [ca. 400 m], 30 Apr. 1932 [l, cap, buds].

Wolf & Everett 9023, CAS, DS, UC, US. Same locality as *Wolf 3165*, 29 July 1937 [l, cap].

SONORA

Gentry 21203, DES, MEXU, US. Km 2509 in mountain pass SE of Quítovac, 4 Sept. 1965, elev. 1,800–2,000 ft [ca. 550–600 m], [l, cap].

Agave gigantensis

BAJA CALIFORNIA

Barclay & Arguelles 1990, MEXU, US. Mountains above Rancho San Sebastián, 28 Apr. 1966.

Carter & Reese 4552, UC. Cuesta de las Parras just above Rancho de las Parras, road between Loreto & San Javier, 5 June 1963, alt. ca. 350 m.

Gentry 7693, 7713 [pars.] ARIZ, DES, MEXU, SD, UC, UM. Picachos de Santa Clara, 5–10 Nov. 1947 [l, f].

Gentry 10339, 10342, DES, MEXU, US. Picachos de Santa Clara, Vizcaino Desert, 4–5 Apr. 1951 [doubtfully assigned here].

Gentry & Arguelles 10327, 10324, MEXU, US. Between Llano San Julio & Sierra de la Giganta, 2 Apr. 1951 [l, f, photo].

Gentry & Fox 11778, DES, MEXU, US. Sierra de las Palmas above Rancho San Sebastián, 27–29 Apr. 1952 [l, cap].

Gentry & McGill 23320, DES, MEXU, **US**. Type. Sierra de las Palmas, above Rancho San Sebastián, 20 June 1973, elev. ca. 4,000 ft. [ca. 1,200 m].

Johnston 3843, CAS, US. Puerto Escondido, 29 May 1921 [l, f], "single plant in wash."

Agave margaritae

BAJA CALIFORNIA

Beauchamp 2149, SD. Magdalena Island, on slopes along arroyos N of Punta Magdalena, elev. ca. 5 m, Apr. 1971 [l, cap].

Beauchamp 2109, SD. West end of N side of Margarita Island, elev. ca. 50 m, 6 Apr. 1971.

Brandegee s.n., **UC**. Type. Magdalena Island, 14 Jan. 1889 [l, f, cap].

Gentry, Fox, Arguelles 11903, 11905, DES, US. Cienegita, Isla Santa Margarita, on sandy bajada or plain, 15 May 1952.

Moran 3540, BH. Santa Maria Bay, 31 Mar. 1952 [l, f].

Moran 4187, BH. Man-of-War Cove, Magdalena Bay, 21 May 1952 [l, cap].

Rose 16261, US. Santa Maria Bay, Isla Santa Magdalena, 18 Mar. 1911 [type of *Agave connochaetodon* Trel.].

Agave mckelveyana

ARIZONA

Bezy 286, ARIZ. 8.3 miles [ca. 13 km] W of Hillside, [Yavapai Co.], elev. 3,900 ft [ca. 1,200 m], 8 June 1964 [f].

Bruem s.n., DS. Dean Mountain Rd., Hualapai Mts., 6 Oct. 1935 [l, caps].

Breitung 18156, DS. McClout Mts., ca. 18 miles [ca. 13 km] N of Congress, [Yavapai Co.], 23 Aug. 1959.

Eastwood 18387, CAS. Aquarius Mts. [Mohave Co.], 13 May 1931 [l, bud].

Eastwood s.n., CAS. Coyote Pass between Oatman and Kingman, May 1931.

Gentry 21979, ARIZ, DES, **US**. Type. Sitgreave Pass in Black Mts., ca. 4 miles [ca. 6 km] NE of Oatman, 26 June 1966.

Gentry 23000, DES, MEXU, US. 26 miles [ca. 42 km] SE of Wikieup along Route 93, Mohave Co., elev. 3,400 ft [ca. 1,040 m], 10 June 1972 [l, f].

Gentry 23002, DES, MEXU, US. 6 miles [ca. 10 km] E of Route 93 along road to Burro Creek, Mohave Co., elev. 3,600 ft [ca. 1,000 m], 10 June 1972.

Gentry & Ogden 9981, DES, MEXU, US. Near Oatman, elev. 3,000–3,200 ft [ca. 900–975 m], low bush cover with scattered juniper and yucca in igneous mountains, [Mohave Co.], 15 Nov. 1950 [l, cap].

Gentry & Whitehead 22312, ARIZ, DES, US. Hualapai Mts., along road to Recreation Park from Kingman, elev. ca. 5,000 ft [ca. 1,500 m], 1 July 1967 [l, first f].

Harrison & Kearney 7303, ARIZ. Table Top Mt., Mohave Co., 16 Aug. 1930 [l, cap].

Haskell, s.n., ARIZ. Near Gold Road, [Mohave Co.], 6 Dec. 1941 [l, cap].

Jones 25167, POM. Near Oatman, [Mohave Co.], 1930 [in part, cap, mixed with leaf of *A. utahensis*].

Kearney & Peebles 12570, ARIZ. Rocky hills S of Hillside, Yavapai Co., elev. 3,600 ft [ca. 1,100 m], 17 Sept. 1935, "scape 8 ft. high (other plants with scapes twice as tall, but leaves always small)" [l, cap].

McKelvey 2235, 1515, 1653, A. West of Burro Creek, Aquarius Mts., [Mohave Co.], 14 May 1931 [l, f, cap].

McKelvey 2250, A. Black Mts., between Kingman and Oatman, [Mohave Co.], 16 May 1931 [f in fluid].

McKelvey 4056A, A. Camp Verde region, 23 Apr. 1934 [l, cap].

McKelvey 4078, 4080, A. Juniper Mts. [Yavapai Co.], Apr. 1934.

McKelvey 1648, 1649, 4071, A. Between Kirkland and Hillside, [Yavapai Co.], 29 Mar. 1930 [whole plant with cap].

Stephens s.n., UC. Hualapai Mts., 5 July 1903 [l, f].

Agave moranii

BAJA CALIFORNIA

Chambers 629, DS. Canyon del Diablo to N and W of Picacho del Diablo, E flank of Sierra San Pedro Mártir, 6 miles [ca. 10 km] from Canyon mouth, elev. 4,500 ft [ca. 1,400 m], 17 June 1954 [f, cap, bract].

Gentry & McGill 23287, DES, MEXU, **US**. Type. 2–3 miles [ca. 3–5 km] SE of Agua Caliente, on E plain of Sierra San Pedro Mártir, 13 June 1973 [l, f].

Moran 18295, DES, MEXU, US. SE side of San Felipe Valley [Sierra Santa Rosa?] elev. ca. 530 m, 9 Mar. 1971.

Moran Photo, SD. Cerro Chato, S side of Sierra San Pedro Mártir, elev. ca. 1,800 m, 3 June 1963.

Moran 21562, SD. Sierra San Pedro Mártir Arroyo del Cajón ca. 2 miles [ca. 3 km] from the mouth, near 30°51′ N, 115°16′ W, elev. ca. 810 m, 28 Dec. 1974.

Agave promontorii

BAJA CALIFORNIA

Barclay & Arguelles 1986, DES, MEXU, US Western summit of Sierra Laguna, 15 Apr. 1966 [l, f].

Brandegee s.n., UC. Sierra de la Laguna, 24 Apr. 1892 [l, f].

Brandegee s.n., UC. San José del Cabo, cultivated in San Diego, 1903.

Gentry 10164, DES, MEXU, US. Huntington Botanical Gardens, San Marino, California, 9–15 Jan. 1951.

Gentry 11218, DES, MEXU, US. Rancho Laguna and vicinity, Sierra Laguna, Cape District western summit, 3 Oct. 1951.

Gentry 11229, DES, MEXU, US. Rancho Burrera at west base of Sierra Laguna, 1–4 Oct 1951.

Gentry 19671, DES. Sierra Laguna, Cape District, 1952 [photo].

Moran 7451, CAS, SD. La Aguja, elev. 1,900 m, Cape District, 18 May 1959 [l, f, cap].

Nelson & Goldman 7437, **US.** Type. From San Bernardo to El Sauz, Sierra La Laguna, 21 Jan 1906, alt. 2,400–5,000 ft [ca. 700–1,500 m], [l, f photo].

Agave sebastiana

BAJA CALIFORNIA

Anthony 264, DS, K, US. San Benito Island Mar.–June 1897 [l, f, fine-toothed form].

Anthony s.n., CAS. San Benito Island, Mar.–June 1897 [f].

Beauchamp 2095, SD. West San Benito Island, 4 Apr. 1971, elev. ca. 130 m.

Beauchamp 3193, SD. Isla San Benito Oeste 28 Feb. 1972. Common on slopes, floral stem 4 feet high [f].

Belding s.n., CAS. May 1881 [fl].

Benedict s.n., SBBG. Cedros Island, upper Campo Punta, Norte Canyon, easterly from Cerro del Norte, 15 Mar. 1971.

Benedict s.n., SBBG. West San Benito Island, 200 yards [ca. 180 m] S of fishing village, SE portion of island, 9 Mar. 1971 [side flowering shoot].

Brandegee s.n., UC. San Benito Island, 27 Mar. 1897 [l, f, br].

Brandegee s.n., UC. San Benito Island, 1 Apr. 1897 [l, f, br].

***Greene s.n.*, CAS,** DS, UC. Type. Cedros Island, 1 May 1885. [leaves 25–30 cm, lanceolate, broadest in middle with regular teeth mostly curved upwards, 2 cm apart, 5–7 mm long, spine slender 20–30 mm long, openly grooved above, flowers 65–67 mm long, 3 l, fl].

Howell 10691, CAS. SE side, Cedros Island, 16 Aug. 1932.

Mason 1986, CAS, K, US. Cedros Island, 3 June 1925 [f, cap].

Moran 2954, CAS. Cedros Island.

Moran 4198, BH, DS. West San Benito Island, 24 May 1952 [l, f—leaf linear-ovate with close-set, slender teeth, reduced below mid-blade, on continuous margin].

Moran 15142, SD. Natividad Island, near middle, elev. ca. 100 m, 24 June 1968 [l, fl]. "Floral stem 1 m high, plant smaller than usual. Fls. bright yellow, erect, visited by flies."

Moran 17430, 17431, DES, SD. West San Benito Island, elev. ca. 100 m, 19 Apr. 1970 [l, fl].

Moran 17449, SD. East San Benito Island, elev. ca. 25 m, 20 Apr. 1970 [l, fl]. Leaves gray to gray-green.

Moran 21206, DES, SD. Cedros Island, 2 miles [ca. 3 km] from N end, elev. ca. 600 m, 28 Mar. 1974 [l, fl]. "Fl. stem 3 m, with 10 branches in upper 3 dm."

Moran 19924, SD. East shore of Bahía Tortugas, elev. ca. 10 m, 8 Feb. 1973 [l, fl].

Philbrick B75–31, SBBG. West San Benito Island, 20 Jan. 1975. W facing slope, halfway between terrace & summit [l, br, bud].

Philbrick & Benedict B72–77, SBBG. Arroyo Madrid, above Playa Madrid, N of Colorada, SE portion of Cedros Island, 20 Apr. 1972 [l, fl].

Solis s.n., MEXU. Isla Cedros, 3 June 1925 [l, f].

Agave shawii goldmaniana

BAJA CALIFORNIA

Barclay & Arguelles 1985, MEXU, US. Ca. 25 miles [ca. 40 km] N of Punta Prieta, 6 Apr. 1966 [l, fl].

Ferris 8581, DS. W of San Borjas Range (Sierra Calamajué), 12 miles [ca. 19 km] from Laguna Chapala along road to Punta Prieta, 6 Mar. 1934 [l, f, caps].

Gentry 10349, MEXU, US. Near Marmolito, 7 Apr. 1951.

Gentry 10361, 10365, 10366, 10355, MEXU, US. Ca. 7 miles [ca. 11 km] S of Tinaja Yubay and 15 miles [ca. 24 km] NE of Punta Prieta, 8 Apr. 1951 [l, f, cap].

Gentry 10379, 10381, 10382, DES, MEXU, US. Six miles [ca. 10 km] N of Socorro, 11 Apr. 1951 [l, fl].

Gentry 11318, 11319, 11320, MEXU, US. Ca. 10 miles [ca. 16 km] N of Punta Prieta, 24 Oct. 1951 [anal., photo].

Gentry 11935, MEXU, US. Ca. 5 miles [ca. 8 km] NW of Mezquitál, Vizcaino Desert, 18 May 1952 [l, f, inflo, photo].

Gentry 19974, MEXU, US. Valley W of Sierra Calamajué, 25 miles [ca. 40 km] N of Punta Prieta, 30 Apr. 1963 [l, f, photo].

Gentry & Fox 11948, DES, MEXU, MICH, US. Ca. 15 miles [ca. 24 km] NE of Punta Prieta, 20 May 1952 [l, f, cap].

Gentry & Gentry 23161, DES, MEXU, US. 16 miles [ca. 26 km] NE of Punta Prieta, 5 Apr. 1973 [fl].

Gentry & Gentry 23166, DES, MEXU, US. 8 miles [ca. 13 km] W of Mission San Borja, 7 Apr. 1973 [l, f, photo].

Harbison s.n., CAS. Near Tinaja Yubay, 30 Apr. 1964 [l, fl].

Harbison s.n., SD. Four miles [ca. 6 km] S of San Andrés, 25 July 1941 [l, fl].

Harbison s.n., SD. Punta Prieta, 29 Apr. 1940 [l, fl].

Hastings & Turner 63-231, DS. Arroyo Aguajito, 14.9 miles [ca. 24 km] E of Rosario, elev. 700 ft [ca. 200 m], 19 Oct. 1963 [l, f, cap].

Lindsay DES 42, DES. Near San Andrés, 1930's [l, f, cap].

Moran 17027, DES, SD. 7.6 miles [ca. 12 km] N of Puerto Santa Catarina, 28 Mar. 1970; "Only this one plant seen in the Santa Catarina drainage" [l, f, cap].

Moran 17053, SD. 6 miles [ca. 10 km] E of Punta Canoas, elev. ca. 50 m, 29 Mar. 1970, "Common but only one still flowering" [l, f, cap].

Moran 17121, SD. Puerto San José, elev. ca. 25 m, 30 Mar. 1970 [l, f].

Moran 17191, SD. 8 miles [ca. 13 km] S of Las Palomas, elev. ca. 140 m, 1 Apr. 1970, "Rosettes 13 dm wide, of 125 leaves, 45 × 12 cm" [l, f].

Moran 17204, SD. Boca de Marrón, elev. ca. 5 m, 2 Apr. 1970, "Common on rocky hillside. Rosettes 12–14 dm wide of 200 leaves" [l, f].

***Nelson & Goldman 7151*, US.** Type. Yubay, 30 miles [ca. 48 km] SE of Calamajué, B.C., alt. 2,000 ft [ca. 600 m], 18 Sept. 1905 [l, cap, photo].

Wiggins 4475, DS. Between El Marmol and Rosario, 40 miles [ca. 64 km] E of Rosario, 12 Mar. 1930 [l, f].

Wiggins & Thomas 171, DS, US. About 14 miles [ca. 23 km] toward the coast from Cerro Blanco (SE of Rosario), alt. ca. 1,200 ft. [ca. 350 m], 8 Feb. 1962 [l, f].

Agave shawii shawii

BAJA CALIFORNIA

Arnott 12, 21, UC. 4 miles [ca. 6 km] N of Socorro, 21 Apr. 1955 [f].

Brandegee s.n., UC. Colnett, May 1893 [l, f].

Cox s.n., UC. San Telmo de Abajo, May 1931.

Farmer s.n., SD. Santo Tomás Valley, 24 Dec. 1955 [series of leaves].

Ferris 8524, DS. 16 miles [ca. 26 km] from Colnett Wash on Santo Domingo Road, 2 Mar. 1934 [l, f].

Ferris 8528, 8529, DS, US. 30 miles [ca. 48 km] N of Rosario on road to Santo Domingo, 3 Mar. 1934.

Gentry 4001, UC. Between San Vicente and Hamilton Ranch, 10 Nov. 1938 [l, f].

Gentry 10079, MEXU, US. Huntington Botanical Gardens, 9–15 Jan. 1951 [l, f].

Gentry 10281, MEXU, US. Km 57 S of Tijuana, 21 Mar. 1951 [l, f].

Gentry 10285, MEXU, US. Ca. 10 miles [ca. 16 km] NE of San Telmo, 23 Mar. 1951 [l, f].

Gentry & Arguelles 19968, MEXU, US. 15 miles [ca. 24 km] E of Rosario, 28 Apr. 1963 [f].

Gentry & Gentry 23154, DES, MEXU, US. 12–13 miles [ca. 19–21 km] E of Rosario, 4 Apr. 1973 [l, f, photo].

Gentry & Gentry 23190, DES, MEXU, US. Near La Misión, 27 miles [ca. 43 km] N of Ensenada by power station, old road, 15 Apr. 1973 [l, f, photo].

Harbison 45522, SD. Santa Maria Valley, 31 Aug. 1953.

Harbison s.n., SD. 7 miles [ca. 11 km] SE San Quintín, 21 Apr. 1927 [l, f].

Harbison s.n., SD. San Simon (ca. 10 miles [ca. 16 km] E of San Quintín), 10 Sept. 1955.

Harbison s.n., SD. Camalu Point, 29 Dec. 1949 [l, f, cap June 1939].

Harbison s.n., SD. Arroyo ESE of Rosario 29 Dec. 1949 [leaves narrow with very fine or small teeth. 1–2 m long & very narrow corneous margin].

Harbison s.n., SD. Canyon above Hamilton Ranch, 15 Dec. 1953 [l, f].

Harbison s.n., SD. La Misión Point, 30 Dec. 1949 [l, f].

Harbison & Howe s.n., SD. Arroyo 8.4 miles [ca. 14 km] E of Rosario, elev. ca. 200 m, 2? Sept. 1965.

Harding s.n., SD. Halfway between Ensenada and Tijuana, 28 Dec. 1937 [br, f].

Jones s.n., DS, MEXU. West of Tijuana near the sea, 26 Dec. 1924 [f].

Moran 16711, DES, SD, UC. Jatay, (S of La Misión), elev. ca. 100 m [f], "Abundant on coastal terrace 1 mile S of Jatay."

Wiggins 21, 442, DS. Valle de San Telmo, 1? miles [ca. 24 km] E of Hwy. No. 1, 17 June 197? [l, f, insects & hummingbirds].

Wiggins & Gillespie 3911, 4006, CAS, DS MEXU, US. 37 miles [ca. 60 km] S of Tijuana 2 miles [ca. 3 km] S of "Halfway House," ? Sept. 1929.

CALIFORNIA

Eastwood 2940, CAS. Boundary Monument San Diego Co., 24 Apr. 1913 [l, f, topotype].

Gander 494, SD. Near Boundary Monument 258, on shore of Pacific, San Diego Co., 28 Jan 1936 [l, f].

Hall 3957, UC. SW corner of San Diego Co. Calif., 30 Apr. 1903 [cap].

Harbison s.n., DES. International Monument, San Diego Co., Calif., [topotype].

***Hitchcock in 1875*, MO.** Type. Initial Boundary Monument, S of San Diego.

Moran 16707, DES, SD, UC. West side of Point Loma [f], 100 m S of Cal-Western Cam pus, 18 Dec. 1969. "Perhaps 200 rosettes, brink of sea bluffs."

Moran 21699, SD. Torrey Pines, San Diego Co., 29 Mar. 1975; "uniform spination—may be one clone—it can be considered native."

Stover s.n., SD. Point Loma, San Diego Co., 21 Apr. 1937 [l, f].

Wolf 5412, CAS. 8 ft. N of Mexican Boundary and second small hill from ocean, San Diego Co., elev. 30–40 ft [ca. 9–12 m], 31 Aug. 1933, topotype.

Wolf 2660, DS. Rancho Santa Ana Botanical Garden from plant brought from SW San Diego Co., 15 Feb. 1932 [l, f, br].

Agave sobria frailensis

BAJA CALIFORNIA

***Gentry & Cech 11264*, MEXU, US.** Type. 4 miles [ca. 6 km] northward of Punta Frailes, 7 Oct. 1951 [l, f].

Gentry & Cech 11257, MEXU, US. 5–8 miles [ca. 8–13 km] N of Punta Frailes, 7 Oct. 1951, [l, photo].

Gentry & Fox 11858, MEXU, US. 5–8 miles [ca. 8–13 km] N of Punta Frailes, 6 May 1952 [l, f, cap].

Agave sobria roseana

BAJA CALIFORNIA

Brandegee s.n., UC. La Paz, 14 Apr. 1892.

Gentry 11274, MEXU, US. Islote Gallo, 10 Oct. 1951 [l, f, cap].

Gentry & Cech 11277, MEXU, US. A W ridge of Isla Espíritu Santo, Golfo de California, 10 Oct. 1951 [l, photo].

Gentry & Fox 11869, MEXU, US. Ca. 3 miles [ca. 5 km] E of La Paz, 7 May 1952 [l, f, photo].

Hastings & Turner 64–154b, SD. 6 miles [ca. 10 km] N of La Paz.

Johnston 3989, 4001, UC, US. The Isthmus, Espiritu Santo Island, 31 May 1921.

Johnston 4002, 4003, UC, US. San Gabriel Bay, Espiritu Santo Island, 1 June 1921 [l, f].

***Rose 16854*, US.** Type. Espiritu Santo Island, 18 Apr. 1911.

Wiggins 17828, DS. Isla Partida, just N of Isla Espíritu Santo, 20 Apr. 1962.

Agave sobria sobria

BAJA CALIFORNIA

Barclay & Arguelles 1989, MEXU, US. Vicinity of San Miguel de Comondú, 26 Apr. 1966, topotype [l, f].

Brandegee s.n., DS, UC. Type. Comondú Mesas, 23 Mar. 1889 [l, margins, br, cap, f].

Brandegee s.n., MO, UC. Cape region mountains, 20 Sept. 1899 [f, type of *A. brandegeei*, mixed with *A. aurea*].

Carter 5486, UC. Cumbre de la Cuesta de Las Parras, north of road, alt. ca. 1,750 m, road from Loreto to San Javier, 5 July 1970.

Carter 5487, UC. Cuesta S of Arroyo Liguí, south of Loreto, 8 July 1970.

Carter & Sharsmith 4940, UC. Along trail from San José de Agua Verde to Bahía Agua Verde, on Gulf drainage, alt. ca. 360 m, 4 June 1965.

Gentry 10304, MEXU, US. Arroyo Purísima several miles above Purísima, 31 Mar. 1951.

Gentry 10308, MEXU, US. Arroyo Purísima above Purísima, 1 Apr. 1951.

Gentry 11303, DES. Ca. 10 miles [ca. 16 km] W of Canipolé [photo].

Gentry 12382, MEXU, US. Ca. 3 miles [ca. 5 km] N of San Javier, Sierra Giganta, 2 Dec. 1952.

Gentry 12387, MEXU, US. E side of Bahía Concepción, Rancho Salto, 3 Dec. 1952.

Gentry & Cech 11291, DES, MEXU, US. Comondú, N slope of volcanic rim, 15 Oct. 1951 [l, photo].

Gentry & Fox 11811, MEXU, US. Rancho San Andreas, Sierra de las Palmas, S of Santa Rosalía, 27–29 Apr. 1952.

Gentry et al. 11876, MEXU, US. 2–3 miles [ca. 3–5 km] NE of La Paz, 7 May 1952.

Gentry et al. 11882, DES, MEXU, US. Comondú, 10 May 1952 [cap].

Harbison s.n., SD. Danzante Island, 7 Apr. 1962 [l, f, cap, very long ovaries].

Harbison s.n., SD. Top of grade on road from Comondú to Loreto, 6 Oct. 1967.

Hutchison 7399, 7473, SD. 13 miles [ca. 21 km] S of El Coyote, Concepcion Bay (½ mile N of km 85) May & Jan. 1975.

Johnston 3857, CAS, US. Ballenas Bay, Danzante Island, 24 May 1921 [l, f, cap—leaf and its armor similar to *A. roseana*].

Johnston 3887, CAS, UC, US. Agua Verde Bay, 26 May 1921.

Moran 3936, BH, DS. Canyon S of Balandra Bay, Carmen Island, 18 Apr. 1952 [l, f].

Purpus s.n., MO, UC. San José del Cabo, Jan.—Mar. 1901 [f only, mixed with *A. aurea*].

Wiggins 11446, UC. 6 miles [ca. 10 km] W of Canipolé, 17 Nov. 1946 [l, different form].

Agave subsimplex

SONORA

Felger 2721, ARIZ. Isla Cholludo (Roca Fuca, Seal Island), 30 Oct. 1958, "common and widespread, suckering, leaves bluish; inflo. branched, only one plant in fl.; anthers yellow, filaments and tepals dusty rose-colored; base of tube pinkish white" [l, f].

Felger & Bezy 14171, ARIZ. Foothills at N side of Cerro Tepopa, 4 miles [ca. 6 km] by road W of 9 miles [ca. 14 km] by road S of Desemboque, 15 May 1966 [f].

Felger & Bezy 14182, ARIZ. 6.3 miles [ca. 10 km] S of Desemboque San Ignacio, middle bajada with sandy soil, 14 May 1966, "few scattered colonies, leaves faintly cross-banded, tepals white to pinkish" [f].

Gentry 4486, DES, ARIZ. Puerto Libertad, 22 May 1939 [f].

Gentry 10217, DES, MEXU, US. N end of Sierra Coloral, 28 Feb. 1951 [l, cap, pl].

Gentry 10221, 10222, 10223, DES, MEXU, US. Cerro Punta Tepopa, 1 Mar. 1939 [l, cap, pl].

Hastings et al. 63–8, ARIZ. Punto Cirio, about 7 miles [ca. 11 km] S of Puerto Libertad, 3 Jan. 1963 [l, cap—small leaf with big teats].

MacDougal & Shreve s.n., ARIZ. Kino Point, 17 Nov. 1923.

Moran 13022, SD, UC. Turner's Island, elev. ca. 25 m, "common on rocky slopes." Tepals light yellow or reddish, filaments pink to dark red, 25 Apr. 1966 [l, f].

***Rose 16811*, US.** Type. Seal Island, just off Tiburon Island, 13 Apr. 1911.

Whiting 9047, ARIZ. Tiburon Island, vicinity of Tecomate, 14–22 June 1951, "abundant on rocky mountain slopes on W side of island. This one from NW corner."

Agave vizcainoensis

BAJA CALIFORNIA

***Gentry 7469*, ARIZ, DES, DS, MEXU, MICH, UC.** Type. Cerro Tordillo, Sierra Vizcaíno, 12–13 Mar. 1947, elev. 400–800 ft [ca. 120–240 m], [l, f].

Gentry 7713, 7693 ARIZ, DES, DS, MICH, UC, UM. Picachos de Santa Clara, Vizcaíno Desert, 5–10 Nov. 1947 [l, cap. Doubtfully assigned here, in part.].

Howell 10660, CAS. San Bartolomé Bay (= Bahía Tortuga), 14 Aug. 1932 [l, f, s].

GLOSSARY OF SPECIAL TERMS

abaxial, the side of a lateral organ away from the central axis.

acaulescent, stemless or without visible stem below the leaves.

acicular, needle-shaped.

adaxial, the side of a lateral organ next to the central axis.

adventitious buds, those produced in areas without visible bud initials, as from the stem instead of the axils of the leaves.

allopatric, applied to allied species or populations inhabiting separate geographic areas. Compare sympatric.

antrorse, directed forward or towards the apex, as the prickles on a leaf margin.

arcuate, moderately arched or curving.

ascending, of leaves pointed upward and outward at about 20° or more from the horizontal.

attenuate, gradually narrowed or prolonged.

bud printing, of leaves when the margin of one leaf is impressed upon the surface of the next leaf.

bulbiferous, producing bulbils.

bulbil, small plant reproduced vegetatively in the axils of the inflorescence. A form of asexual reproduction.

cabeza, Spanish, head, applied to the thick, short stem of agaves.

campanulate, bell-shaped.

castaneous, chestnut-colored.

caulescent, having a stem or trunk below the leaves.

cespitose, as applied to succulents, growing in clusters by the production of basal branches, suckers, or offsets.

chartaceous, papery, dry, and thin.

circadian (L. *circa*, about, and *dies*, day), relating to biologic variations or rhythms with a cycle of about 24 hours.

clone, a group of individual plants reproduced asexually from a single original parent.

conduplicate, folded together lengthwise.

conic, cone-shaped.

contingent perennial, a plant living more than 2 years and whose flowering is contingent upon the proper climatic conditions, i.e., rainfall and higher temperature.

crenate, applied to leaf margin strongly and abruptly undulate with large teats.

cucullate, hooded at the apex.

cultigen, a plant known to exist only in cultivation. Compare cultivate.

cultivate, as a noun, a cultivated plant with known wild ancestors.

deflexed, bent downward.

descending, of leaves, directed below the horizontal.

ensiform, sword-shaped.

explanate, flattened, spread out flat.

exserted, exceeding the corolla, as the filaments extending beyond the tepals.

filiferae, threadlike structures along the leaf margins.

filiferous, having threadlike structures.

friable, said of teeth or leaf margins that are easily brushed or rubbed off.

funnelform, funnel-shaped.

glaucous, whitened with a waxy coating over the epidermis.

guttered, having the sides of the leaves raised to form a trough-shaped leaf, partly conduplicate.

half, handle, applied to the narrowed lower part of the leaf, which is usually the least prickly part for grasping with the hand.

keel, the fleshy midrib on the tepal or the leaf.

laterals, the main branches of the paniculate inflorescence.

maguey, an American Indian name for agave. It was picked up by the earliest Spaniards and appears to have been in general use in the Caribbean region and Mexico. Cortes in "Historia de Mexico" wrote, "miel de unas plantas, que llaman en las otras, y estas maguey, que es muy mejor que arrope; y de estas plantas hacen azucar, y vino, que es asi mismo venden."

mescal, an American Indian name applied to agave plants, to the cooked parts of the same, and to the distilled liquor made from the meristem. Used more in northern Mexico. Also applied loosely to *Manfreda*, *Tillandsia*, and other monocots.

metl, a Nahuatl name for agave, still in use among native tribal people in central Mexico; "papalo metl" (butterfly agave, *A. potatorum*), "tlaca metl" (*A. salmiana*).

monocarpic, a plant or rosette that flowers once and dies. Compare polycarpic.

multiannual, a plant that flowers once and dies, but requires several to many years to mature.

neck, apical portion of ovary between the ovarian cells and the base of the tube.

panicle, the branched inflorescence of the subgenus *Agave* with flowers borne in umbellate clusters on lateral branches.

paniculate, like a panicle.

patulous, standing open, spreading.

plane, applied to leaves having the upper surface flattened as compared with guttered or explanate.

polycarpic, a plant or rosette that flowers repeatedly, but not necessarily every year. Compare monocarpic.

proterandrous, the condition of a perfect flower when the anthers dehisce before the pistil is receptive. Hence, the flower cannot normally be self-fertilizing.

pruinose, having a waxy exudate on the surface of the leaf.

pulque, the fermented juice of the larger agaves. Word derived from Nahuatl "poliuhqui" or "ocli poliuhqui," but which was applied to soured or spoiled "ocli," fide Nuñez Ortega, the chronicler of Hernando Cortes.

raceme, an inflorescence in which the flowers are borne on pedicels along a central axis.

reclinate, reclining, applied to leaves pressing downward upon lower leaves or on the ground.

recurved, recurvate, curved backward or downward.

reflexed, bent sharply downward.

retrorse, directed towards the base.

rhizome, underground stem or shoot.

rosette, a closely spaced group of radiating leaves limited to a portion of the stem, usually at the base of the inflorescence.

sapogenin, a compound derived by hydrolysis from saponin. A large group of such compounds have been found in plant tissues and named according to their specific molecular configuration, e.g., diosgenin, smilogenin, hecogenin.

shaft, in *Agave* the central axis of the inflorescence including the peduncle and the central rachis of the flowering portion or branches.

spike, an inflorescence with the flowers more or less sessile along a common or single peduncle or shaft.

spine, terminal spine, in *Agave* the pungent indurated tip of the leaf.

spreading, of leaves extending outward less than 20° from horizontal.

subulate, awl-shaped, long tapering.

surculose, producing suckers or offsets.

sympatric, applied to related species or populations inhabiting the same geographic area.

teat, fleshy prominences under the teeth on the leaf margins.

teeth, the prickles along the leaf edge.

tepal, a combination of sepal and petal, applied when the segments of the perianth are not differentiated into two dissimilar ranks, (common in monocotyledons).

trigonous, three-angled with plane faces.

tubular, tube-shaped.

umbel, a flat-topped or low-rounded flower cluster with the pedicels of unequal length from a common point, like an inverted umbrella.

umbellate, having the inflorescence in umbels or in similar form.

undulate, applied to leaf margins with low teats, wavy, as compared to straight margins.

urceolate, urn-shaped.

valleculate, little valley-shaped, as applied to agaves having folds in the leaf towards apex, plicate.

Literature Cited

Anderson, Edgar. 1949. Introgressive hybridization. John Wiley & Sons, New York. 109 pp.

Aschmann, Homer. 1959. The central desert of California: demography and ecology. Univ. Calif. Press, Berkeley & Los Angeles. 315 pp.

Baegert, Johann Jakob. 1972. Nachrichten von der Amerikanischen halbinsel, Californien. Mannheim. [Also, Spanish translation by Pedro R. Hendricks Pérez, *Noticias de la Peninsula Americana de California por el Rev. Padre Juan Jacobo Baegert. Con una introducción por Paul Kirchoff.* México, Antigua libraría robredo de J. Perrúa e hijos, 1942. A résumé of original published in *Ann. Geogr.* 5:237–239, 1866. An English translation by M. M. Brandenburg & C. L. Baumann entitled *Observations in Lower California*. Berkeley, 1952.]

Barco, Miguel del. 1973. Historia natural y crónica de la Antigua California. [Adiciones y correcciones a la noticia de Miguel Venegas] Edición estudio preliminar, notas y apéndices. By Miguel León-Portilla. Universidad Nacional Autónoma de México, Inst. Invest. Hist. Méx. lxxxv + 464 pp.

Barrows, David Prescott. 1967. The ethno-botany of the Coahuilla Indians of southern California. Including a Coahuilla bibliography and introductory essay by Harry W. Lawton, Lowell John Beand [and] William Bright. Malki Museum Press, Banning, Calif. 82 pp. [Reprint of Ph.D. Thesis, Univ. Chicago, 1897]

Berger, Alwin. 1915. Die Agaven. 285 pp., illus. Jena, Verlag von Gustav Fischer.

Bolton, Herbert Eugene (editor). 1919. Kino's historical memoir of Pimería Alta. Cleveland, 2 vols., illus., maps.

Burrus, Ernest J. (editor and translator). 1966. Wenceslaus Linck's diary of his 1766 expedition to northern Baja California. Dawson's Book Shop, Los Angeles. 115 pp.

———. 1967. Wenceslaus Linck's reports and letters, 1762–1778. Dawson's Book Shop, Los Angeles. 94 pp.

Castetter, E. F., W. H. Bell, and A. R. Grove. 1938. The early utilization and distribution of *Agave* in the American Southwest. Univ. Mexico, Bull. 6. 92 pp.

Cave, Marion S.. 1964. Cytological observations on some genera of the Agavaceae. Madroño 17(5):163–170.

Clavigero, Francisco Xaviero. 1789. Storia della California. Venice. 2 vols., map. [Spanish translation by Nicolas García de San Vicente, entitled *Historia de la Antiqua ó Baja California*. Mexico, 1933, 267 pp. English translation by Sara E. Lake & A. A. Gray, *The history of Lower California*. Stanford Univ. Press, Stanford, 1937, 413 pp., maps.]

Darton, N. H. 1921. Geological reconnaissance in Baja California. J. Geol. 29:720–748.

Felger, Richard, and Mary B. Moser. 1970. Seri use of *Agave* (century plant). The Kiva 35:159–167.

Gentry, Howard Scott. 1972. The agave family in Sonora. U.S. Dep. Agric. Agric. Handbk. 399. Pp. 1–195, maps, illus.

———, and Jane R. Sauck. 1978. A study of the stomatal complex in *Agave*: groups Deserticolae, Campaniflorae, Umbelliflorae. Proc. Calif. Acad. Sci., Ser. 4, 41(16):371–387, figs. 1–28, tables 1–2.

Goldman, Edward A. 1951. Biological investigations in Mexico. Smithson. Miscell. Coll. 115:1–476, map, illus.

Granick, Elsa B. 1944. A karyosystematic study of the genus *Agave*. Am. J. Bot. 31:283–289.

Hastings, James R. 1964. Climatological data for Baja California. Univ. Ariz. Inst. Atoms. Physics Tech. Rep. No. 14.

———, and Robert R. Humphrey. 1969. Climatological data and statistics for Sonora and northern Sinaloa. Univ. Ariz. Inst. Atmos. Physics Tech. Rep. No. 19.

Johnston, Ivan M. 1924. Expedition of the California Academy of Sciences to the Gulf of California in 1921. XXX. The botany (the vascular plants). Proc. Calif. Acad. Sci., Ser. 4, 12:951–1218.

Langman, Ida K. 1964. A selected guide to the literature on the flowering plants of Mexico. Univ. Penn. Press, Philadelphia. 1015 pp.

Larson, R. L., H. W. Menard, and S. M. Smith. 1968. Gulf of California: a result of ocean-floor spreading and transform faulting. Science 161:781–783.

Moran, Reid. 1964. Floración de *Agave goldmaniana* a los 31 años. Cact. Succulentas Méx. 9:87–88.

———. 1967. *Agave Subsimplex* Trelease. Cact. Succulentas Méx. 12:59–61.

Nelson, Edward W. 1921. Lower California and its natural resources. Mem. Natl. Acad. Sci. 16(1st Mem.):1–194, illus.

Rau, Charles. 1864. An account of the aboriginal inhabitants of the California Peninsula as given by Jacob Baegert . . . . Annu. Rep. Smithson. Inst. (1863):352–369.

Schuchert, Charles. 1935. Historical geology of the Antillean-Caribbean region. John Wiley & Sons, New York. 811 pp.

Shreve, Forrest. 1951. Vegetation of the Sonoran Desert. Carnegie Inst. Washington Publ. 591. xii + 192 pp., 37 pls.

———, and Ira R. Wiggins. 1964. Vegetation and flora of the Sonoran Desert. Vols. 1 & 2. Stanford Univ. Press, Stanford, Calif. 1740 pp.

Trelease, William. 1912. The agaves of Lower California. Missouri Bot. Gard. Annu. Rep. (1911)22:37–65, pls. 18–72.

Wall, M. E. 1954. Steroidal sapogenins XV. U.S. Dep. Agric. Agric. Res. Serv. AIC-367. 32 pp. (processed)

——— et al. 1954a. Steroidal sapogenins XII. J. Am. Pharm. Assoc. 43:503–505.

——— et al. 1954b. Steroidal sapogenins VII. Survey of plants for steroidal sapogenins and other constituents. J. Am. Pharm. Assoc. (Sci. Ed.) 43:1–7.

——— et al. 1955. Steroidal sapogenins XXVI. U.S. Dep. Agric. Agric. Res. Serv. ARS-73-4. 30 pp. (processed)

——— et al. 1957. Steroidal sapogenins XLIII. J. Am. Pharm. Assoc. (Sci. Ed.) 46:653–684.

Wilson, J. Tuzo, et al. 1972. Continents adrift: readings from *Scientific American*, with introduction by J. Tuzo Wilson. W. H. Freeman & Co., San Francisco. 172 pp.

Wolf, Ivan, et al. 1960–65. U.S. Dep. Agric. Unpubl. Rep. Inter-departmental from North. Utilization Research & Development Division, Peoria, Illinois.

INDEX

Page numbers in bold face indicate taxon description and essay; synonyms marked with an asterisk*.

Spotted something that isn't quite right?

Let us know and we'll fix it for future reprints. And as a thank you for your diligence, we'll send you a free Sothis book of your choice.

publisher@sothis.press

This page intentionally left blank